Lecture Notes in Mathematics

Edited by A. Dold and B. Eckmann

758

C. Năstăsescu
F. Van Oystaeyen

Graded and Filtered Rings and Modules

Springer-Verlag
Berlin Heidelberg New York 1979

Authors

C. Năstăsescu
Faculty of Mathematics
University of Bucharest
Str. Academiei 14
R – 70109 Bucharest 1
Romania

F. Van Oystaeyen
Department of Mathematics
University of Antwerp UIA
Universiteitsplein 1
B – 2610 Wilrijk
Belgium

AMS Subject Classifications (1980): Primary: 16A03
Secondary: 16A05, 16A33, 16A45, 16A50, 16A55, 16A63, 16A66

ISBN 3-540-09708-2 Springer-Verlag Berlin Heidelberg New York
ISBN 0-387-09708-2 Springer-Verlag New York Heidelberg Berlin

Library of Congress Cataloging in Publication Data
Năstăsescu, Constantin.
Graded and filtered rings and modules.
(Lecture notes in mathematics; 758)
Bibliography: p.
Includes index.
1. Associative rings. 2. Graded rings. 3. Graded modules. 4. Filtered rings. 5. Filtered modules. I. Oystaeyen, F. van, 1947- joint author. II. Title. III. Series: Lecture notes in mathematics (Berlin); 758.
QA3.L28 no. 758 [QA251.5] 510'.8s [512'.4] 79-26585
ISBN 0-387-09708-2

Printing and binding: Beltz Offsetdruck, Hemsbach/Bergstr.
2141/3140-543210

PREFACE

The aim of these lecture notes is to present coherently the latest and less known results concerning the theory of filtered and graded rings and modules. With an eye to projective algebraic geometry, the theory of commutative graded rings and Noetherian graded modules over them has been expounded in the large, e.g. using cohomological methods and Serre's theorem, which states that the category of coherent sheaves over projective n-space is equivalent to a category of graded Noetherian modules over a graded polynomial ring, it is possible to prove generalizations of Max Noether's and Bezout's theorem within the latter category, cf. [15].
In writing these notes we have adopted a completely different point of view, i.e. a purely ring theoretical one, and the only reference to projective geometry will be in II.14, II.15 and II.16 but then over a non-commutative ring. The material presented falls naturally into three parts.

Chapter I is devoted to the study of the relation between a filtered ring R and the category of filtered R-modules on one hand, and the graded ring and module category associated to R on the other hand. In particular we focussed attention on : Krull dimension in Gabriel's and Rentschler's sense; free, projective and finitely generated objects, homological dimension i.e. projective and flat dimension etc.... Most of the results included in this chapter have been developed after those included in [35].

In Chapter II we introduce several topics in, what may be called, "Graded Ring Theory". The general problem faced is the following : what information about R-mod, the category of left R-modules, can be drawn from certain properties holding in R-gr, the category of graded left R-modules, i.e. the feed-back from the graded theory into ungraded ring theory. Of course in investigating this feed-back problem it will also be necessary to develop some graded techniques paralleling the ungraded theory. In this perspective we study : finitness conditions, Krull dimension, homological dimension, primary decompositions, graded rings and modules of quotients, etc.... Although the graded version of A. Goldie's theorems does not yield a necessary and sufficient condition for the existence of a graded semisimple gr. Artinian ring of fractions, under some conditions on the gradation of the ring, which may be rather easily verified in real life circumstances, our graded Goldie rings do have the desired rings of fractions. Part of the results in Chapter II have been obtained by us in recent papers, cf. [28], [29], [31], but we have also included several generalizations and new results.

Chapter III contains some new results in the theory of Noetherian graded commutative rings on a problem put forward by Nagata in [26]. This problem has been solved by several authors amongst whom we quote [8], [23], [24]. However, the presentation of the problem given in this chapter is by far more simple than the one given in the papers cited above; moreover our solution to the problem allows a number of generalizations as well as some new results. Actually part of Chapter III may be generalized to the case of a non-commutative ring, using the localization techniques of Chapter II, but we thought that full generality everywhere would have been disturbingly incongruent with respect to the aim of the chapter and so we presented only the commutative theory.

A certain knowledge of the elements of Ring Theory and of Homological Algebra, on the level of graduate courses, is necessary for the understanding of these notes; in other words, we assume some familiarity with concepts like : Ext^n, the Jacobson radical of a ring, Grothendieck categories, but the indispensable background material on filtrations and gradations has been included in the notes.

C. Năstăsescu F. Van Oystaeyen

Acknowledgement

Each of the authors expresses gratitude towards the department of mathematics of the co-author's institution for its hospitality while these notes were being conceived and written.

We thank Prof. Dr. P.M. Cohn for reading the manuscript and for making some useful suggestions.

Our typist, Ludwig Callaerts, nearly got sick turning our illegible notes into this nicely typed book; we really appreciate his work and thank him a lot

F. Van Oystaeyen thanks Danielle for her patience and support, while he was working at home, doing things that could have been done during normal working hours.

CONTENTS

SOME NOTATIONS AND CONVENTIONS

All rings have identity elements. All R-modules are left R-modules unless otherwise stated. Ideal of R will mean two-sided ideal of R.

References without chapter indication refer within the chapter containing the reference.

$\mathbb{N}$	:	the natural numbers $\{0,1,2,\dots\}$
$\mathbb{Z}$	:	the integers
R-mod	:	the category of left R-modules
R-gr	:	the category of graded left R-modules
$\underline{\ }$	:	if $M \in$ R-gr then $\underline{M}$ is the underlying R-module
$L_g(R)$	:	the lattice of graded left ideals of R
FM	:	if $M \in$ R-filt then FM is the filtration $\{F_iM,\ i \in \mathbb{Z}\}$
$\hat{M}$	:	if $M \in$ R-filt then $\hat{M}$ is the completion of M
T_n	:	the suspension functor on R-gr
Ext_R^n	:	the n-th derived functor of $\mathrm{Hom}_R(-,-)$
R_+	:	$R_+ = \bigoplus_{i>0} R_i = R_{>0}$
G	:	the „associated gradation" functor : R-filt $\to$ G(R)-gr
R^+	:	$R^+ = \bigoplus_{i\geq 0} R_i = R_{\geq 0}$
M^-	:	$M^- = \bigoplus_{i\leq 0} M_i = M_{\leq 0}$
$X^\sim$	:	the graded module generated by homogeneous elements of maximal degree in the decompositions of elements of X.
$X_\sim$	:	the graded module generated by homogeneous elements of minimal degree in the decompositions of elements of X.
rad	:	the prime radical
$(M)_g$	:	If M is an R-submodule contained in a graded module N then $(M)_g$ is the graded submodule of N maximal with the property of being contained in M.
$M^{(d,k)}$	:	if $(d,k) \in \mathbb{Z}^2$ such that $d \geq 1$, $0 \leq k \leq d-1$ then $M^{(d,k)} = \bigoplus_{i\in\mathbb{Z}} M_{id+k}$
$B^g(M)$	:	the graded bi-endomorphism ring of M
J(R)	:	the Jacobson radical of R

$J^g(R)$	:	the graded Jacobson radical of R
$Ann_R M$	:	the annihilator of M in R.
$M_n(R)$	:	the n by n matrices over R
ht(P)	:	the height of P
Ass M	:	the set of prime ideals associated to M
x^*	:	the external homogenization in M[T] of $x \in M$
$\mathcal{L}(\kappa)$	:	the filter associated to the kernel functor κ
Q^g_κ	:	the graded localization functor associated to κ
$Q_{\underline{\kappa}}$	:	the localization functor in R-mod associated to $\underline{\kappa}$
G(P)	:	Goldie's multiplicative set associated to the prime ideal P
C.M.	:	Cohen-Macaulay
Proj R	:	the projective spectrum of R i.e. the set of graded prime ideals not containing R_+.
p.dim	:	the projective dimension of M
inj.dim M	:	the injective dimension of M
w.dim M	:	the weak (flat) dimension of M
gl.dim R	:	the left global dimension of R
gl.w.dim R	:	the left weak global dimension of R

All these dimensions have analogues in R-gr denoted by gr.p.dim M, gr.inj.dim M, gr.w.dim M, gr.gl.dim R, gr.gl.w.dim R.

CHAPTER I. FILTERED RINGS AND MODULES.

I.1. PRELIMINARIES.

1.1. Definition. An associative ring R with unit is said to be a filtered ring if there is an ascending chain $\{F_nR, n \in \mathbb{Z},\}$ of additive subgroups of R satisfying the following conditions : $1 \in F_oR$, $F_nR.F_mR \subset F_{n+m}R$ for any $(n,m) \in \mathbb{Z}^2$. The family of these subgroups, denoted FR is called the filtration of R.

Note that the definition implies that F_oR is a subring of R.

1.2. Examples.

1°. Any ring R can be made into a filtered ring by means of the trivial filtration, i.e. $F_nR = o$ if $n < o$, $F_nR = R$ if $n \geq o$.

2°. Let I be an ideal of R. The I-adic filtration is obtained by putting $F_nR = R$ for $n \geq o$ while $F_nR = I^{-n}$ for $n < o$.

3°. Let R be any ring and let $\varphi : R \to R$ be an injective ring homomorphism, $\delta : R \to R$ a φ-derivation i.e. a group homomorphism for the additive structure of R such that $\delta(xy) = \delta(x)y + x^{\varphi}\delta(y)$ for $x,y \in R$. The ring of skew polynomials $R[X,\varphi,\delta]$ is obtained from R by adjoining a variable X and defining multiplication according to : $Xb = r^{\varphi}X + \delta(r)$, for $r \in R$, in such a way that R becomes a subring of $R[X,\varphi,\delta]$. (Note that the action of endomorphisms is written exponent-wise). The fact that φ is injective allows to introduce the degree function (denoted by : deg) in the usual way and so we may define the degree-filtration by : $F_nR[X,\varphi,\delta] = \{P \in R[X,\varphi,\delta], \deg P \leq n\}$; hence $F_oR[X,\varphi,\delta] = R$.

1.3. Definition. Let R be a filtered ring. A left R-module M is called a filtered module if there exists an ascending chain $\{F_nM, n \in \mathbb{Z}\}$ of additive subgroups of M such that $F_nR.F_mM \subset F_{n+m}M$ for any $(n,m) \in \mathbb{Z}^2$. The family $\{F_nM, n \in \mathbb{Z}\}$ is the filtration of M. Obviously, if R is a filtered ring then ${}_RR$ (resp. R_R) is a filtered left (resp. right) R-module.

If M is a filtered left R-module then its filtration F may have one of the following properties :

(E) F is exhaustive if $M = \bigcup_{n \in \mathbb{Z}} F_n M$.

(D) F is discrete if there is an $n_o \in \mathbb{Z}$ such that $F_i M = o$ for all $i < n_o$.

(S) F is separated if $\bigcap_{n \in \mathbb{Z}} F_n M = o$.

(C) F is complete if $M = \varprojlim_n M/F_{-n}M$.

If the filtration of R is trivial and $M \in R$-mod, then any ascending chain of submodules of M defines a filtration for M; in this case the trivial filtration of M is given by $F_n M = o$ if $n < o$, $F_n M = M$ if $n \geq o$.

1.4. Definition. Let R and S be filtered rings an $M \in R$-mod-S is said to be a filtered R-S-module if there exists an ascending chain $\{F_n M, n \in \mathbb{Z}\}$ of additive subgroups such that : $F_n R . F_m M \subset F_{m+n} M$ for all $n,m \in \mathbb{Z}$, and $F_m M . F_t S \subset F_{m+t} M$ for all $m,t \in \mathbb{Z}$.

I.2. THE CATEGORY OF FILTERED MODULES.

Let R be filtered ring, M and N filtered left R-modules. An R-homomorphism $f \in \mathrm{Hom}_R(M,N)$ is said to have degree p if $f(F_i M) \subset F_{i+p} N$ for all $i \in \mathbb{Z}$. Homomorphisms of finite degree from a subgroup $\mathrm{HOM}_R(M,N)$ of $\mathrm{Hom}_R(M,N)$, homomorphisms of degree p form a subgroup $F_p \mathrm{HOM}_R(M,N)$ of $\mathrm{HOM}_R(M,N)$,

One easily checks :

1. If $p \leq q$ then : $F_p \mathrm{HOM}_R(M,N) \subset F_q \mathrm{HOM}_R(M,N)$.
2. $\mathrm{HOM}_R(M,N) = \bigcup_{p \in \mathbb{Z}} F_p \mathrm{HOM}_R(M,N)$.
3. If $f : M \to N$ has degree p and $g : N \to P$ has degree q then gof has degree p+q.

These properties allow us to introduce the category R-filt of filtered left R-modules where the morphisms are the homomorphisms in $\mathrm{HOM}_R(-,-)$ of degree o. If $M,N \in R$-filt then $F_o \mathrm{HOM}_R(M,N)$ will simply be denoted by $\mathrm{Hom}_{FR}(M,N)$. Clearly R-filt is an additive category and furthermore if $f \in \mathrm{Hom}_{FR}(M,N)$ then Ker f and Coker f are in R-filt. Indeed, Ker f has the induced filtration $F_i \mathrm{Ker}\, f = \mathrm{Ker}\, f \cap F_i M$, whereas Coker f is filtered by putting $F_i\, \mathrm{Coker}\, f = (\mathrm{Im} f + F_i N)/\mathrm{Im} f$. In particular it follows that monomorphisms and epimorphisms in R-filt coincide with the injective resp. surjective morphisms of R-filt.

Obviously, arbitrary direct sums as well as direct products exist in R-filt. Moreover, if (M_i, φ_{ij}), with i,j in some index set I, is an inductive (projective) system in R-filt then its inductive (projective) limit exists in R-filt; it is the R-module $\varinjlim M_i$ ($\varprojlim M_i$) with filtration $F_p \varinjlim M_i = \varinjlim_i F_p M_i$, ($F_p \varprojlim M_i = \varprojlim_i F_p M_i$).

2.1. Remarks.

1. Let $\{M_i, i \in I\}$ be a family of objects from R-filt such that their filtrations FM_i, $i \in I$, are exhaustive, then the filtration of $\bigoplus_I M_i$ is exhaustive. This property fails for the direct product $\prod_I M_i$! The similar result however does hold for inductive or projective limits.

2. R-filt is preabelian (cf. [13]) but not abelian. Indeed, take $M \neq o$ in R-mod and put $F_i M = o$ for all $i \in \mathbb{Z}$. Let F'M be another filtration on M and denote G, resp. H, the filtered modules obtained from M using FM resp. F'M. The identity morphism of M is in $\mathrm{Hom}_{FR}(G,H)$ but not in $\mathrm{Hom}_{FR}(H,G)$ and thus the identity of M is a bijective mapping which fails to be an isomorphism.

3. In a similar way as before one defines the category R-filt-S for a pair of filtered rings R and S.

2.2. Functors on R-filt.

Before studying some special functors in R-filt let us point out that, if R is a filtered ring with filtration FR, we have a functor D : R-mod $\to$ R-filt which associates to $M \in$ R-mod the R-module M with filtration $F_n M = F_n R.M$. This filtration of M is called the deduced filtration. Note that $F_n D(M) = F_n R.M = M$ for any $n \geq o$.

The suspension functor. For $n \in \mathbb{Z}$ we define the n-th suspension functor T_n : R-filt $\to$ R-filt, which associates to $M \in$ R-filt with filtration FM the filtered module M(n) with filtration $F_i M(n) = F_{i+n} M$ hence $F_i T_n(M) = F_{i+n}(M)$. Suspension is characterized by : $T_n \circ T_m = T_{n+m}$, $T_o = \mathrm{Id}$ (the identity functor). In particular, every T_n is an equivalence of categories which commutes with direct sums, products, inductive and projective limits. Consequently $M \in$ R-filt is complete (i.e. FM is complete) if and only if $T_n(M)$ is complete. It is easily verified that the following holds : for $p \in \mathbb{Z}$

and for $M,N \in R\text{-filt}$:

$$F_p HOM_R(M,N) = Hom_{FR}(M(-p),N) = Hom_{FR}(M,N(p)) .$$

The exhaustion functor. Let R have filtration FR. If we put $R' = \bigcup_{n\in\mathbb{Z}} F_nR$ then, R' is a filtered subring of R. If $M \in R\text{-filt}$ then, clearly, $M' = \bigcup_{n\in\mathbb{Z}} F_nM$ is a filtered R'-module and the filtration FM' given by $F_nM' = F_nM$ for all $n \in \mathbb{Z}$ is an exhaustive filtration. In this way we obtain a functor : R-filt → R'-filt carrying M into the corresponding M', which is called the exhaustion functor.

The completion functor. Let $M \in R\text{-filt}$. A sequence $(m_i)_{i\in\mathbb{N}}$ of elements of M is said to be a Cauchy sequence if for every $p \in \mathbb{N}$ there is an $N(p) \in \mathbb{N}$ such that $m_s - m_t \in F_{-p}M$ for all $s,t \geq N(p)$. A sequence $(m_i)_{i\in\mathbb{N}}$ converges to $m \in M$ if for every $p \in \mathbb{N}$ there exists $N(p) \in \mathbb{N}$ such that $m_s - m \in F_{-p}M$ whenever $s \geq N(p)$. It is well known, cf. [3], that M is complete if and only if FM is separated and all Cauchy sequences converge in M. Using this completeness criterion it is straightforward to check that any finite direct sum of complete modules is again a complete module.

Now let $M \in R\text{-filt}$ and consider the projective system : $\{M/F_pM,\ \pi_{pq} : M/F_{-p}M \to M/F_{-q}M \quad p \geq q\}$. Then $\hat{M} = \varprojlim_p M/F_{-p}M$ is a filtered $\mathbb{Z}$-module because of the results in I.2. Moreover, $\hat{M}$ is in this way a complete abelian group and there is a natural isomorphism between $\hat{M}/F_p\hat{M}$ and M/F_pM for every $p \in \mathbb{Z}$. If the filtration of R is exhaustive then $\hat{M}$ is a filtered R-module, in particular $\hat{R}$ is a complete filtered ring. If FM is exhaustive too then $\hat{M}$ has the structure of a filtered $\hat{R}$-module. Thus we obtain a functor $\underline{c}$, called the completion functor, from the category of filtered R-modules with exhaustive filtration to the category of complete filtered $\hat{R}$-modules associating $\hat{M}$ to M.

Taking the limit of the system of the canonical maps $\pi_p : M \to M/F_pM$ yields an R-linear map $c_M : M \to \hat{M}$ and M is complete if and only if c_M is an isomorphism i.e. if $M \cong c(M) = \hat{M}$. Note that the assumption that the filtration considered are exhaustive is not really restricting since we may always apply the exhaustion functor if this were not the case.

I.3. GRADED RINGS AND MODULES.

A ring R is said to be a graded ring of type $\mathbb{Z}$ if there is a family of additive subgroups $\{R_n, n\in\mathbb{Z}\}$ of R such that $R = \bigoplus_{n\in\mathbb{Z}} R_n$ and $R_iR_j \subset R_{i+j}$ for $i,j\in\mathbb{Z}$. It follows from this that $1\in R_0$ and that R_0 is a subring of R. An $M\in R$-mod is said to be a graded left R-module if there is a family $\{M_n, n\in\mathbb{Z}\}$ of additive subgroups of M with the properties : $M = \bigoplus_{n\in\mathbb{Z}} M_n$ and $R_iM_j \subset M_{i+j}$ for all $i,j\in\mathbb{Z}$.
Although for an abelian group G it is very well possible to study gradations of type G we will only consider gradations of type $\mathbb{Z}$ in this paper.
The elements of $h(R) = \bigcup_{n\in\mathbb{Z}} R_n$ and $h(M) = \bigcup_{n\in\mathbb{Z}} M_n$ are called homogeneous elements of R and M resp. If $m\neq o$, $m\in M_i$, then m is called an homogeneous element of degree i, we write : deg m = i. Any nonzero $m\in M$ may be written, in a unique way, as a finite sum $m_1 + \dots + m_k$ where none of the terms is zero and such that $\deg m_1 < \deg m_2 < \dots < \deg m_k$; the elements $m_1,\dots,m_k$ in such an expression are the homogeneous components of m.

3.1. Examples.

1°. Let R be a ring, $\varphi : R\to R$ an injective ring homomorphism. The ring of twisted polynomials $S = R[X,\varphi]$ is a graded ring with grading : $S_i = o$ if $i<o$, $S_i = \{aX^i, a\in R\}$ if $i\geq o$.

2°. Any ring R may be considered as a graded ring with trivial gradation : $R_0 = R$ and $R_i = o$ if $i\neq o$.

3°. Let k be a field, X and Y variables and consider the ring $k(X,Y)_h$ consisting of rational functions $\frac{\varphi(X,Y)}{\varphi(X,Y)}$ where φ and ψ are homogeneous polynomials in X and Y. This is a graded ring with grading induced by the degree function : $\deg(\frac{\varphi(X,Y)}{\psi(X,Y)}) = \deg\varphi(X,Y)$ $\deg\psi(X,Y)$. Commutative graded rings of this type are well known, they arise as homogeneous coordinate rings and function fields of projective curves in Algebraic Geometry.

3.2. The Category of Graded Modules

Let $R = \bigoplus_{i\in\mathbb{Z}} R_i$ be a $\mathbb{Z}$-graded ring and M a $\mathbb{Z}$-graded R-module (we omit $\mathbb{Z}$ in what follows).

A submodule N of M is a graded submodule if $N = \bigoplus_{i\in\mathbb{Z}} (N\cap M_i)$ or, equivalently, if for any $x \in N$ the homogeneous components of x are again in N.

Now consider graded left R-modules M and N. An R-linear $f : M \to N$ is said to be a <u>morphism of degree p</u> if : $f(M_i) \subset N_{i+p}$ for any $i \in \mathbb{Z}$. Morphisms of degree p from an additive subgroup of $\mathrm{Hom}_R(M,N)$, which we will denote by $\mathrm{HOM}_R(M,N)_p$ and it is clear that $\bigcup_{p \in \mathbb{Z}} \mathrm{HOM}_R(M,N)_p = \mathrm{HOM}_R(M,N)$ is a graded abelian group. Composition of a morphism $f : M \to N$, of degree m, with a morphism $g : N \to P$, of degree n, yields a morphism gof of degree m+n.

If N is a graded submodule of M then $M/N = \bigoplus_{i \in \mathbb{Z}} (M_i + N)/N$ is graded and the canonical projection $M \to M/N$ is a graded morphism of degree o.

Now we introduce the category R-gr of graded left R-modules where the morphisms in this category are taken to be the graded morphisms of degree o. This category possesses direct sums, products, inductive and projective limits. One easily verifies this statement for sums and products, furthermore, if $(M_\alpha, f_{\alpha\beta})$ is some inductive system (resp. projective system) of objects in R-gr, then the left module $\varinjlim_\alpha M_\alpha$ (resp. $\varprojlim_\alpha M_\alpha$) may be equipped with the grading $(\varinjlim_\alpha M_\alpha)_n = \varinjlim_\alpha (M_\alpha)_n$, (resp. $(\varprojlim_\alpha M_\alpha)_n = \varprojlim_\alpha (M_\alpha)_n$). If $M, N \in$ R-gr then $\mathrm{Hom}_{R\text{-gr}}(M,N)$ denotes the graded morphisms of degree o from M to N. If $f \in \mathrm{Hom}_{R\text{-gr}}(M,N)$, then both Ker f and Coker f are in R-gr. Indeed, since Im f is a graded submodule of N, it follows that $\mathrm{Coker}\, f = N/\mathrm{Im}\, f = \bigoplus_{i \in \mathbb{Z}} (N_i + \mathrm{Im}\, f)/\mathrm{Im}\, f$, is the cokernel of f in R-gr. It is not hard to check that Im f and Coim f are isomorphic in R-gr, therefore R-gr is an abelian category which satisfies Grothendieck's axioms Ab3, Ab4, Ab3* and Ab4*, cf. [13]. The axiom Ab5 may equally easily be deduced. Obviously we have a functor $U : \text{R-gr} \to \text{R-mod}$ (U for ungrading) which associates to a graded $M \in$ R-gr the corresponding ungraded R-module.

<u>Notation.</u> We will write $U(M) = \underline{M}$, but we omit this for R since it will be clear from the context whether R is being considered as a graded ring or not. One of the main problems concerning R-gr that we will be dealing with may be phrased as follows : <u>if $M \in$ R-gr has property P, then it is true that $\underline{M}$ has this property too</u>. The converse is usually true and far more easier to deal with.

Similar to the definitions in 2.2 we define the <u>n^{th}-suspension</u> M(n) of $M \in$ R-gr to be the module M together with the gradation $M(n)_i = M_{n+i}$. This defines a functor $T_n : R\text{-gr} \to R\text{-gr}$ which may be characterized by :

1°. $T_n \circ T_m = T_{n+m}$.

2°. $T_n \circ T_{-n} = Id.$

3°. $U \circ T_n = U.$

In particular T_n is an equivalence of categories, for all $n \in \mathbb{Z}$. It is straightforward to prove that the family $\{R(n), n \in \mathbb{Z}\}$ is a family of generators for R-gr, consequently $\bigoplus_{n \in \mathbb{Z}} R(n)$ is a generator of R-gr and R-gr is a Grothendieck category with <u>enough injective objects</u>, cf. [13]. An $F \in R\text{-gr}$ is said to be <u>free</u> if it has a basis consisting of homogeneous elements, equivalently, if there is a family $\{n_i, i \in I\}$ of integers such that $F \cong \bigoplus_{i \in I} R(n_i)$. Since any $M \in R\text{-gr}$ is isomorphic to a quotient of a free object of R-gr, whereas free objects are certainly projective in R-gr, it follows that R-gr has <u>enough projective objects</u>.

3.3. Some Properties of Graded Modules.

An $M \in R\text{-gr}$ is <u>left limited</u> (resp. <u>right limited</u>) if there is an $n_o \in \mathbb{Z}$ such that $M_i = o$ for all $i < n_o$, (resp. $i > n_o$). If $i < o$ yields $M_i = o$ then M is said to be <u>positively graded</u>, (negatively graded is defined in the same way).

<u>3.3.1. Lemma.</u> Let R be a left limited graded ring and $M \in R\text{-gr}$, then the following statements hold :

a. If M is finitely generated then M is left limited.

b. If M is left limited then there exists a free graded module F, left limited, and an epimorphism $F \to M \to o$.

c. M is finitely presented in R-gr if and only if $\underline{M}$ is a finitely presented R-module.

<u>Proof</u> : a. Since M is finitely generated there exists a finitely generated free graded module F such that we have an exact sequence in R-gr : $F \to M \to o$. Since R is left limited and F being finitely generated, it follows that F is left limited and so is M.

b. Let $n_o \in \mathbb{Z}$ be such that $R_i = o$ for $i < n_o$ and let $m_o \in \mathbb{Z}$ be such that $M_i = o$ for all $i < m_o$. If $x \neq o$ is in $h(M)$ then $\deg x \geq m_o$. To an $x \in h(M)$ there corresponds a morphism of degree o, $\varphi_x : R(-n_x) \to M$, given by $\varphi_x(1) = x$, where $n_x = \deg x$. Note that for $i < n_o + m_o$ we have $R(-n_x)_i = o$. Putting $\varphi = \underset{x \in h(M)}{\oplus} \varphi_x$ yields an exact sequence :

$$\underset{x \in h(M)}{\oplus} R(-n_x) \xrightarrow{\varphi} M \longrightarrow o$$

where $\underset{x \in h(M)}{\oplus} R(-n_x)$ is clearly a free, left limited graded module.

c. Obvious. □

3.3.2. Lemma. Suppose that $M, N \in R\text{-gr}$ and that M is finitely generated, then : $\mathrm{HOM}_R(M,N) = \mathrm{Hom}_R(\underline{M},\underline{N})$.

Proof : Obviously, graded morphisms are R-linear. Conversely, let $f : M \to N$ be R-linear and let $x_1, \ldots, x_n$ be homogeneous generators for M. Express $f(x_j), j = 1 \ldots n$, as $\sum_{i=1}^{k} y_{ji}$ with $\deg y_{j1} < \ldots < \deg y_{jk}$. Each element y_{ji} defines a graded morphism g_{ji}, $g_{ji} : M \to N$, in the following way : $g_{ji}(x_k) = o$ if $k \neq j$, $g_{ji}(x_j) = y_{ji}$ (it is easily checked that this is well-defined). Now putting $g = \sum_{i=1}^{k} g_{ji}$ yields $f = g$ and therefore $f \in \mathrm{HOM}_R(M,N)$. □

3.3.3. Remark. If M is not finitely generated then it can happen that $\mathrm{HOM}_R(M,N) \neq \mathrm{Hom}_R(\underline{M},\underline{N})$. For example let R have a non-limited grading, then there exists an element $(a_n)_n \in \prod_{n \in \mathbb{Z}} R_n$ which is not in $\underset{n \in \mathbb{Z}}{\oplus} R_n$. Put $M = R^{(\mathbb{Z})}$ and define $f \in \mathrm{Hom}_R(M,R)$ by putting $f((x_n)_n) = \sum_{i \in \mathbb{Z}} a_i x_i$. Clearly, $f \notin \mathrm{HOM}_R(M,R)$!

The derived functors of the left exact functor $\mathrm{HOM}_R(.,.)$ will be denoted by $\mathrm{EXT}_R^n(.,.)$, then we have

3.3.4. Corollary. If R is a graded left Noetherian ring and M is a finitely generated graded R-module then, for any $n \geq o$: $\mathrm{EXT}_R^n(M,N) = \mathrm{Ext}_R^n(M,N)$ for any graded R-module N.

Proof. Since R is left Noetherian, M has a finite resolution

$$\ldots \longrightarrow F_2 \longrightarrow F_1 \longrightarrow F_o \longrightarrow M \longrightarrow o$$

where the F_i are free graded R-modules and also finitely generated. Now Lemma 3.3.2. finishes the proof. □

Now consider a positively graded ring R and consider the graded ideal $R_+ = \bigoplus_{i>o} R_i$ of R.

<u>3.3.5. Lemma.</u> Let R be a positively graded ring, let $M \in R$-gr be left limited, then $R_+M = M$ if and only if $M = o$.

<u>Proof.</u> Suppose $M \neq o$ and pick $n_o \in \mathbb{Z}$ such that $M_i = o$ for all $i < n_o$, whereas $M_{n_o} \neq o$. Clearly $M_{n_o} \cap R_+M = o$ and therefore $R_+M \neq M$. □

<u>3.3.6. Lemma.</u> Let R be a graded ring, $M,N,P \in R$-gr, and consider the following commutative diagram of R-morphisms :

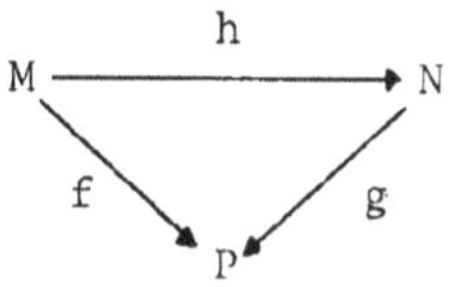

where f has degree o. If g (resp. h) has degree o then there exists a graded morphism h' (resp. g') having degree o such that $f = g \circ h'$ (resp. $h = g' \circ h$).

<u>Proof.</u> Let us prove the case where g has degree o. Take $x \in M_n$ and put $h'(x) = h(x)_n$. It is not hard to see that $f = g \circ h'$. □

<u>3.3.7. Corollary.</u> Let $P \in R$-gr, then P is a projective object in R-gr if and only if $\underline{P}$ is a projective R-module.

<u>Proof.</u> If P is projective in R-gr then it is a direct summand of a free graded object F but then it is also true that $\underline{P}$ is projective in R-mod. Conversely, suppose that $\underline{P}$ is projective in R-mod. Since $P \in R$-gr there exists a free graded R-module F such that $F \xrightarrow{f} P \to o$ is exact in R-gr. Projectivity of $\underline{P}$ entails that there is an R-morphism $g : \underline{P} \to \underline{F}$ such that $f \circ g = 1_P$. According to Lemma 3.3.6. there exists a

graded morphism $g' : P \to F$, of degree o and such that $f \circ g' = 1_P$. Consequently P is a projective object in R-gr. □

3.3.8. Corollary. Let $M \in$ R-gr and N a graded submodule of M. Then N is a direct summand of M if and only if $\underline{N}$ is a direct summand of $\underline{M}$.

A graded left R-submodule I of R is called an homogeneous or graded left ideal of R, if I is moreover two-sided then it is called an homogeneous or graded ideal of R.

3.3.9. Lemma. Let R be a graded ring, $E \in$ R-gr. The following assertions are equivalent :

1. E is an injective object of R-gr.
2. The functor $HOM_R(-,E)$ is exact.
3. For every homogeneous left ideal I of R, the morphism :
   $HOM_R(i,1_E) : HOM_R(R,E) \to HOM_R(I,E)$ deriving from the canonical inclusion $i : I \to R$, is surjective.

Proof. $1 \Leftrightarrow 2$. Obviously E is an injective object if and only if the functor $Hom_{R\text{-}gr}(-,E) = HOM_R(-,E)_0$ is exact. Moreover E is injective if and only if E(n) is injective for every $n \in \mathbb{Z}$.

$2 \Rightarrow 3$. Obvious.

$3 \Rightarrow 1$. The proof is similar to the proof of Baer's theorem in the ungraded case. □

3.3.10. Corollary. If $E \in$ R-gr is such that $\underline{E}$ is an injective R-module then E is injective in R-gr.

3.3.11. Remarks.

1°. If $E \in$ R-gr is injective then $\underline{E}$ need not be injective. For example, let K be any field and consider $R = K[X,X^{-1}]$ with gradation $R_i = \{aX^i, a \in K\}$, $i \in \mathbb{Z}$. It may be checked that R is injective in R-gr but not in R-mod. More about this kind of objects will be in .

2°. If $L \in$ R-gr is such that $\underline{L}$ is a free left R-module then L need not be free in R-gr.

For example, let R be $\mathbb{Z}\times\mathbb{Z}$ with trivial grading. Let L be $\underline{R}$ endowed with the following grading : $L_0 = \mathbb{Z}\times\{o\}$, $L_1 = \{o\}\times\mathbb{Z}$ and $L_i = o$ for $i \neq o,1$.
Obviously, L is not free in R-gr.

The projective dimension of an R-module $\underline{M}$ will be denoted by $p.\dim_{\underline{R}}\underline{M}$. If $M \in$ R-gr, then the projective dimension of M in the category R-gr will be denoted by $gr.p.\dim_R M$. One easily deduces from 3.3.7. that :

3.3.12. Corollary. If $M \in$ R-gr then $gr.p.\dim_R M = p.\dim_{\underline{R}}\underline{M}$.

For further use we state :

3.3.13. Lemma. Let $M \in$ R-gr and let N be a graded submodule of M. If N is essential in M in the graded sense, then $\underline{N}$ is essential in $\underline{M}$.

Proof. Recall that a subobject of M is said to be essential if its intersection with all non-zero subobjects of M is nonzero. Hence, the fact that N is R-gr-essential in M means that, for each $x \in h(M)-\{o\}$ there exists $a \in h(R)$ such that $ax \in N-\{o\}$. If $x \neq o$ is in M then we may write $x = x_{i_1} + \ldots + x_{i_n}$, where the x_{i_k} are the homogeneous components of x of degree i_k. The proof goes by induction on n, establishing that for each $x \in \underline{M}$ there is an $a \in R$ such that $ax \in \underline{N}$. For $n = 1$, this is obvious from graded essentiality of N in M. Define $y = x_{i_1} + \ldots + x_{i_{n-1}}$. Since $x_{i_n} \neq o$ there is an $a \in h(R)$ such that $ax_{i_n} \in N$ and $a\,x_{i_n} \neq o$. Hence $ax - ax_{i_n} = ay$ where ay has at most n-1 homogeneous components. So if $ay = o$ then $ax \in N$ and we are done. If $ay \neq o$ then by induction, we may take some $b \in R$ such that $bay \in N$ and $bay \neq o$. Thus, $bax = bax_{i_n} + bay \in N$. Now $bax = o$ yields $bay = bax_{i_n} = o$, contradicting the choice of b, therefore $bax \neq o$. □

3.3.14. Remark. Let $M \in$ R-gr and let N be a small graded submodule of M (small is the notion dual to essential), then $\underline{N}$ need not be small in $\underline{M}$. For example, consider the graded ring $R = K[X]$ over a field K. Then the homogeneous ideal (X) is small in the graded sense but not in R-mod.

3.4. Tensor Products of Graded Modules.

When talking about $\mathbb{Z}$-gr we assume that $\mathbb{Z}$ is equipped with the trivial grading (it is the only possible grading on $\mathbb{Z}$).

3.4.1. Definition. Let R be a graded ring, $M \in$ gr-R and $N \in$ R-gr. Consider the abelian group $\underline{M \otimes_R N}$ and define its grading by putting $(M \otimes_R N)_n$ equal to the additive subgroup generated by elements $x \otimes y$ with $x \in M_i, y \in N_j$, such that $i+j = n$. In this way we obtain an object $M \otimes_R N$ of $\mathbb{Z}$-gr, called the tensor-product of the graded modules M and N.

Let R^o be the opposite ring of R i.e. the same abelian group but with multiplication reversed. From the above definition it is clear that we obtain a functor $\otimes_R : R^o\text{-gr} \times R\text{-gr} \to \mathbb{Z}\text{-gr}$. Fixing $M \in$ R-gr, then the functor $\otimes_R M : R^o\text{-gr} \to \mathbb{Z}\text{-gr}$ is right exact. M is said to be gr-flat if $\otimes_R M$ is exact. Let us mention some elementary properties, the proof of which is similar to the corresponding proof in the ungraded case.

3.4.2. For any $m,n \in \mathbb{Z} : M(m) \otimes_R N(n) = (M \otimes_R N)(m+n)$. If R and S are graded rings, $M \in$ gr-R, $N \in$ R-gr-S, then $M \otimes_R N$ is a graded right S-module.

3.4.3. If $M \in$ gr-R, $P \in$ gr-S, $N \in$ R-gr-S, then there is a natural isomorphism :

$$\mathrm{HOM}_S(M \otimes_R N, P) \cong \mathrm{HOM}_R(M, \mathrm{HOM}_S(N,P)).$$

3.4.4. If $P \in$ gr-R, $N \in$ S-gr, $M \in$ S-gr-R, then there exists a canonical morphism :

$$\phi : P \otimes_R \mathrm{HOM}_S(M,N) \to \mathrm{HOM}_S(\mathrm{HOM}_R(P,M), N),$$

defined by $\phi(p \otimes f)(g) = (f \circ g)(p)$ for $p \in P$, $f \in \mathrm{HOM}_S(M,N)$ and $g \in \mathrm{HOM}_R(P,M)$.

3.4.5. Proposition. An $M \in$ R-gr is gr-flat if and only if $\underline{M}$ is flat in R-mod.

Proof. If $\underline{M}$ is flat in R-mod then M is obviously gr-flat. Conversely, as in [21], it is easy to show that M is the inductive limit in R-gr of free graded modules, consequently $\underline{M}$ will be flat in R-mod. □

3.4.6. Corollary. The gr-flat dimension of M, denoted by : gr-w $\dim_R M$, is defined as in the ungraded case, (in R-mod, the flat dimension of $\underline{M}$ is denoted by $w.\dim_R M$). The above proposition implies : $\text{gr-w.}\dim_R M = w.\dim_R \underline{M}$.

Let $o \to M' \to M \to M'' \to o$ be exact in R-gr. This sequence is said to be gr-pure if for any $N \in$ gr-R, $o \to N \otimes_R M' \to N \otimes_R M \to N \otimes_R M'' \to o$ is exact in $\mathbb{Z}$-gr .

3.4.7. Corollary. An exact sequence $o \to M' \to M \to M'' \to o$ in R-gr is gr-pure if and only if $o \to M' \to M \to M'' \to o$ is pure in R-mod.

Proof. The idea of the proof for ungraded modules together with Theorem 2.3. of [21] allow to deduce that if $o \to M' \to M \to M'' \to o$ is gr-pure, then it is an inductive limit of a filtered set of exact splitting sequences of the form :

$$o \longrightarrow M' \longrightarrow M' \oplus P^{(i)} \longrightarrow P^{(i)} \longrightarrow o \ ,$$

where each $P^{(i)}$ is of finite presentation. Consequently the sequence is pure in R-mod and so is $o \to \underline{M}' \to \underline{M} \to M'' \to o$. The converse is obvious. □

I.4. Filtration and Associated Gradation.

Let R be a filtered ring, $M \in$ R-filt. Consider the abelian groups : $G(R) = \bigoplus_{i \in \mathbb{Z}} F_iR/F_{i-1}R$, $G(M) = \bigoplus_{i \in \mathbb{Z}} F_iM/F_{i-1}M$. If $x \in F_pM$ then x_p denotes the image of x in $G(M)_p = F_pM/F_{p-1}M$. If $a \in F_iR$, $x \in F_jM$ then define $a_i.x_j = (ax)_{i+j}$ and extend it to a $\mathbb{Z}$-bilinear mapping $\mu : G(R) \times G(M) \to G(M)$. Taking $M = R$, μ makes G(R) into a graded ring and in general G(M) is thus made into a graded G(R)-module. Let $f \in \mathrm{Hom}_{FR}(M,N)$ for some $M,N \in$ R-filt; then f induces canonical mappings $f_i : F_iM/F_{i-1}M \to F_iN/F_{i-1}N$, given by $f_i(x_i) = (f(x))_i$ for $x \in F_iM$, each $i \in \mathbb{Z}$. Putting $G(f) = \bigoplus_{i \in \mathbb{Z}} f_i$ defines a morphism of G(R)-modules.

All of this amounts to the statement that $G :$ R-filt $\to$ G(R)-gr is a functor. We have the following :

4.1. Proposition. The functor G has the following properties :

1. If $M \in$ R-filt and FM is exhaustive and separated then $M = o$ if and only if $G(M) = o$

2. If $n \in \mathbb{Z}$ and $M \in R$-filt, then $G(T_n M) = T_n(G(M))$.

3. If $M \in R$-filt and FM is discrete then $G(M)$ is left limited.

4. If $M \in R$-filt and both FR and FM are exhaustive then $G(M) = G(\hat{M})$.

5. The functor G commutes with direct sums, products and inductive limits.

Proof. Assertions 1., 2., 3., are clear. Statement 4. is well known, cf. [3] chapter III. In order to prove statement 5, let $(M_\alpha)_{\alpha \in J}$ be an arbitrary family of objects of R-filt, and put $M = \bigoplus_{\alpha \in J} M_\alpha$. We have :

$$G(M)_n = F_n(M)/F_{n-1}(M) = \bigoplus_{\alpha \in J} F_n(M_\alpha)/ \bigoplus_{\alpha \in J} F_{n-1}(M_\alpha)$$

$$= \bigoplus_{\alpha \in J} F_n M_\alpha / F_{n-1} M_\alpha = \bigoplus_{\alpha \in J} G(M_\alpha)_n$$

Hence $G(M) = \bigoplus_{\alpha \in J} G(M_\alpha)$. In a similar way we may establish that G commutes with direct products and inductive limits. □

4.2. Remark. If R is a graded ring and $M \in R$-gr then we can define an exhaustive and separated filtration on R (resp. M) by means of the subgroups $F_p R = \bigoplus_{i \leq p} R_i$, (resp. $F_p M = \bigoplus_{i \leq p} M_i$). Let us denote the obtained filtered ring (resp. filtered module) by R' (resp. M'). It is straightforward to prove that $G(R') \cong R$, $G(M') \cong M$. Similarly, the subgroups $F'_p R = \bigoplus_{i \geq -p} R_i$ (resp. $F'_p M = \bigoplus_{i \geq -p} M_i$) define a filtration on R (resp. M). Denoting by R'' (resp. M'') the obtained filtered ring (resp. module) then again $G(R'') \cong R$ and $G(M'') \cong M$.

In studying the effect of G on morphisms we need :

4.3. Definition. Let $M, N \in R$-filt. A filtered morphism $f : M \to N$ is said to be strict if $f(F_p M) = \mathrm{Im} f \cap F_p N$ for each $p \in \mathbb{Z}$.
A sequence $L \xrightarrow{f} M \xrightarrow{g} N$ in R-filt is strict exact if it is an exact sequence in R-mod such that both f and g are strict morphisms in R-filt.

4.4. Theorem. Let R be a filtered ring and consider the following o-sequence in R-filt :

$$(\star) \qquad L \xrightarrow{f} M \xrightarrow{g} N ,$$

as well as the sequence $G(\star)$ in $G(R)$-gr :

$$G(L) \xrightarrow{G(f)} G(M) \xrightarrow{G(g)} G(N).$$

Then :

1. If $(\star)$ is strict exact then $G(\star)$ is exact.
2. If $G(\star)$ is exact and FM is exhaustive then g is strict.
3. If $G(\star)$ is exact while FL is complete and FM is separated then f is strict.
4. If $G(\star)$ is exact and FM is discrete then f is strict.
5. If FK is complete, FM is exhaustive and separated, or if FM is exhaustive and discrete then $(\star)$ is strict exact if and only if $G(\star)$ is exact.

Proof. 1. Clearly $G(g) \circ G(f) = o$. Look at an $x \in F_pM$ which is such that $G(g)x_p = o$. This says $(g(x))_p = o$ and therefore $g(x) \in F_{p-1}N$. Now since g is strict there must exist an $x' \in F_{p-1}M$ such that $g(x) = g(x')$ i.e. $x - x' = f(y)$ with $y \in F_pL$, but then $G(f)(y_p) = x_p$ i.e. $\mathrm{Im}\, G(f) = \mathrm{Ker}\, G(g)$.

2. Let $y \in F_pN \cap \mathrm{Im}\, g$ and $y \notin F_{p-1}N$. There is an $x \in M$ such that $g(x) = y$ and since FM is exhaustive we may assume that $x \in F_{p+s}M$ for some $s \geq o$. If $s = o$ then we are done. If $s > o$ then $G(g)(x_{p+s}) = o$ and the fact that $G(\star)$ is exact implies that $x_{p+s} = G(f)(z_{p+s})$ for some $z \in F_{p+s}L$. Thus $x - f(z) \in F_{p+s-1}M$ and $y = g(x) = g(x - f(z)) = g(x')$ with $x' \in F_{p+s-1}M$. Repeating this procedure we finally get that there is an $u \in F_pM$ such that $y = g(u)$.

3. Take $y \in F_pM \cap \mathrm{Im}\, f$. Exactness of $G(\star)$ yields : $G(g)(y_p) = g(y_p) = o$, therefore $y_p = G(f)(x_p^{(p)})$ for some $x^{(p)} \in F_pL$. Hence $y - f(x^{(p)}) \in \mathrm{Im} f \cap F_{p-1}M$. By induction we find a sequence $x^{(p)}, x^{(p-1)}, \dots, x^{(p-s)}$ with $x^{(p-s)} \in F_{p-s}L$ such that :

$$y - f(x^{(p)}) - \dots - f(x^{(p-s)}) \in \mathrm{Im} f \cap F_{p-s-1}M.$$

Completeness of FL allows to define $x = \sum_{s=o}^{\infty} x^{(p-s)} \in F_pL$. Hence $y - f(x) = y - \lim_{o \to \infty} f(x^{(p)} + x^{(p-1)} + \dots + x^{(p-s)}) = o$, the latter since FM is separated, and it follows from this that $y \in f(F_pL)$. That $f(F_pL) \subset F_pM \cap \mathrm{Im} f$ is obvious.

4. Along the lines of 3.

5. Strict exactness of (*) implies exactness of $G(*)$ because of 1. Conversely, suppose that $G(*)$ is exact and let $y \in M$, $y \neq o$ be such that $g(y) = o$. Since FM is exhaustive we have $y \in F_pM$ and $y \notin F_{p-1}$ for some $p \in \mathbb{Z}$. We obtain thus : $G(g)(y_p) = o$ and so : $y_p = G(f)(x_p^{(p)}) = f(x^{(p)})_p$ for some $x^{(p)} \in F_pL$. Consequently $y - f(x^{(p)}) \in F_{p-1}M$. By induction we obtain $x^{(p)}, \ldots, x^{(p-s)}$ with $x^{(p-s)} \in F_{p-s}L$ and such that :

$$y - f(x^{(p)} + \ldots + x^{(p-s)}) \in F_{p-s-1}\, M \quad .$$

If FM is discrete then for some index s, $F_{p-s-1}M = o$; hence $y = f(x^{(p)} + \ldots + x^{(p-s)})$ and thus (*) is exact. If FM is complete then we may take $x = \sum_{s=o}^{\infty} x^{(p-s)}$ and get $y = f(x)$. The fact that f and g are strict morphisms will follow from 2. and 3..

<u>4.5. Corollary</u>. Let $f : M \to N$ be a morphism in R-filt and suppose that FM and FN are separated and exhaustive, while FM is also complete. If $G(f)$ is bijective then f is bijective too.

<u>4.6. Remark</u>. The restrictions on the filtrations i.e. being exhaustive and complete may be keyed down a lot in most situations because the functor G commutes with the exhaustion and the completion functor so that we have a commutative diagram of functors : $(G(R) = G'(R') = \hat{G}(\hat{R}'))$

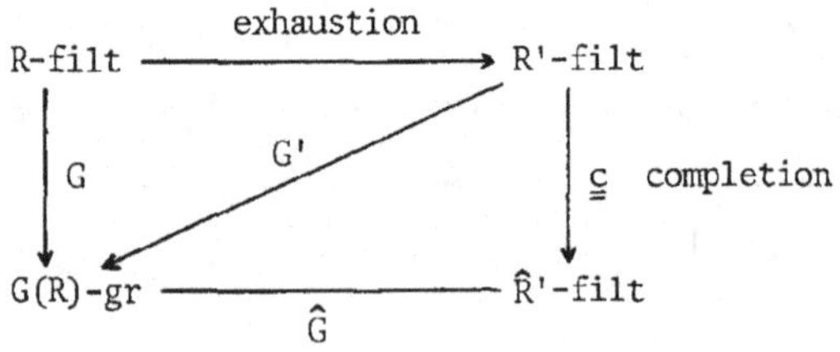

<u>I.5. Free, Projective and Finitely Generated Objects of R-filt.</u>

Let R be a filtered ring, then $L \in$ R-filt is <u>filt-free</u> if it is free in R-mod and has a basis $(x_i)_{i \in J}$ consisting of elements with the property that there exists a family $(n_i)_{i \in J}$ of integers such that :

$$F_pL = \sum_{i\in J} F_{p-n_i} R\,.\,x_i = \bigoplus_{i\in J} F_{p-n_i} R\,.\,x_i \ .$$

Note that for any $i\in J$, $x_i \in F_{n_i}L$ and $x_i \notin F_{n_i-1}L$. We say that $(x_i,n_i)_{i\in J}$ is a filt-basis for L. Next lemma follows easily from this definition and Theorem 4.4.

5.1. Lemma. Let R be a filtered ring, $L\in R$-filt, then :

1°. L is filt-free with filt-basis $(x_i,n_i)_{i\in J}$ if and only if $L \cong \bigoplus_{i\in J} R(-n_i)$.

2°. If L is filt-free with filt-basis $(x_i,n_i)_{i\in J}$ then G(L) is a free graded module in G(R)-gr with homogeneous basis $\{(x_i)_{n_i}, i\in J\}$.

3°. If G(L) is a free object in G(R)-gr with homogeneous basis $\{(x_i)_{n_i}, i\in J\}$ and if FL is discrete, then L is filt-free in R-filt with filt-basis $(x_i,n_i)_{i\in J}$.

4°. If $M\in G(R)$-gr is free graded then there exists a filt-free $L\in R$-filt such that $G(L)\cong M$.

5°. Let $L\in R$-filt be filt-free with filt-basis $(x_i,n_i)_{i\in J}$, let $M\in R$-filt. If $f:\{x_i, i\in J\}\to M$ is a function such that $f(x_i)\in F_{s+n_i}M$, then there is a unique filtered morphism of degree s, $g:L\to M$ which extends f.

6°. Let $M\in R$-filt and suppose that L is filt-free. If $g:G(L)\to G(M)$ is a graded morphism of degree s then there is a filtered morphism $f:L\to M$ of degree s such that $G(f)=g$.

7°. Let $L\in R$-filt be filt-free, then FL is exhaustive (separated) if and only if FR is exhaustive (separated).

If FR is discrete and $\{n_i \in \mathbb{Z}, i\in J\}$ is bounded below then FL is discrete. If J is finite and FR is complete then FL is complete.

8°. Let $M\in R$-filt be such that FM is exhaustive. Then there is a free resolution of M in R-filt :

$$(\star)\quad \ldots \longrightarrow L_2 \xrightarrow{f_2} L_1 \xrightarrow{f_1} L_o \xrightarrow{f_o} M \longrightarrow o$$

where every L_j is filt-free and every f_j is a strict morphism. Moreover, if FR and FM are discrete then we may assume that every FL_j, $j\geq o$, is also discrete.

9°. If FR is exhaustive and complete and if G(R) is left Noetherian and G(M) being finitely generated as a left graded R-module then we may select the L_j in the resolution (*) to be finitely generated too.

Note that an $M \in R$-filt is said to be finitely generated or filt-f.g., if there is a finite family $(x_i,n_i)_{i\in J}$ with $x_i \in M$ and $n_i \in \mathbb{Z}$, such that $F_pM = \Sigma_{i=1}^n F_{p-n_i} R.x_i$

5.2. Remarks. 1°. If $L \in R$-filt is filt-free as well as finitely generated in R-mod then R is filt-f.g.
2°. If $M \in R$-filt is such that FM is exhaustive then M is filt-f.g. if and only if there exists a filt-free L which is finitely generated, and a strict epimorphism $\pi : L \to M$.

5.3. Proposition. Let R be a filtered complete ring and let $M \in R$-filt be such that FM is separated and exhaustive. Then M is filt-f.g. if and only if G(M) is finitely generated as a graded G(R)-module. Furthermore, if G(M) may be generated by n homogeneous elements, then M may be generated as an R-module by less than n generators.

Proof. From the remark in 5.2.2° it follows that, if M is filt-f.g., then G(M) is finitely generated. Conversely let G(M) be generated by homogeneous generators $x^{(i)}_{p_i}$, $1 \leq i \leq n$.
Consider $L \in R$-filt, L filt-free with filt-basis $(y_i,p_i)_i$. Define $f : L \to M$ by $f(y_i) = x^{(i)}$. Since $G(f) : G(L) \to G(M)$ is an epimorphism, Theorem 4.4.5 yields that f is a strict epimorphism and then Remark 5.2.2° entails that M is filt-f.g. and generated by $x^{(i)}$ as an R-module. □

Proposition 5.3 has some consequences for the determination of the Krull dimension of certain modules. In [34] Gabriel and Rentschler defined the Krull dimension of ordered sets for finite ordinal numbers, in [16] this definition has been extended by G. Krause for some other ordinal numbers. Let $(E,\leq)$ be an ordered set. If $a,b \in E$ we put : $[a,b] = \{x \in E,\ a \leq x \leq b\}$, $\Gamma(E) = \{(a,b), a \leq b\}$. By transfinite recurrence, we define on $\Gamma(E)$ the following filtration : $\Gamma_{-1}(E) = \{(a,b),\ a = b\}$, $\Gamma_0(E) = \{(a,b) \in \Gamma(E),\ [a,b]$ is Artinian$\}$, supposing $\Gamma_\beta(E)$ has been defined for all

$\beta < \alpha$, put :

$\Gamma_\alpha(E)=\{(a,b)\in\Gamma(E), \forall b\geq b_1\geq \dots \geq b_n \geq \dots \geq a$ there is an $n\in\mathbb{N}$ such that $[b_{i+1},b_i]\in\Gamma_\beta(E)$ for all $i\geq n\}$.

We obtain an ascending chain : $\Gamma_{-1}(E)\subset\Gamma_0(E)\subset\dots\subset\Gamma_\alpha(E)\subset\dots$. There exists an ordinal ξ such that $\Gamma_\xi(E)=\Gamma_{\xi+1}(E)=\dots$. In case there is an ordinal α such that $\Gamma(E)=\Gamma_\alpha(E)$, E is said to have Krull dimension. The smallest ordinal with the property that $\Gamma_\alpha(E)=\Gamma(E)$ will be called the Krull dimension of E and we denote it by K-dim E.

5.4. Lemma. Let E,F be two ordered sets and let $f : E\to F$ be a strictly increasing mapping. If F has Krull dimension then E has Krull dimension and K dim $E\leq$ K dim F. (cf. [34])

5.5. Lemma. Let E,F be ordered sets with Krull dimension, then $E\times F$ has Krull dimension and K dim $E\times F$ = sup(K.dim E,K dim F). (cf. Note that $E\times F$ has the product ordering.). If A is an arbitrary abelian category, $M\in A$, then we consider the set E of all subobjects of M ordered by inclusion. If E has Krull dimension then M is said to have Krull dimension and we denote it by : $\text{K.dim}_A M$ or simply K.dim M if no confusion is possible. Let $M\in A$ have K.dim $M=\alpha$; then M is said to be α-critical if K.dim $M/M'<\alpha$ for every nonzero subobject M' of M. For example M is o-critical if and only if M is a simple object in the category A. Also it is obvious that if $M\in A$ is α-critical then any nonzero subobject of M is α-critical.
If A is R-mod, $M\in$ R-mod, and if M has Krull dimension in the above sense then we denote it by $\text{K-dim}_R M$. The Krull dimension of $R\in$ R-mod is called the left dimension of R; it is denoted by K-dim R.

Returning to the consequences of Proposition 5.3 we have :

5.6. Corollary. Let $M\in$ R-filt, such that FM is separated and exhaustive and suppose that FR is complete. If G(M) is left Noetherian as an object of G(R)-gr then M is left Noetherian too. Moreover, we have then that : $\text{K-dim}_R M\leq \text{K-dim}_{G(R)} G(M)$.
If $\text{K-dim}_R M=\text{K-dim}_{G(R)} G(M)=\alpha$ and G(M) is α-critical, then M is an α-critical R-module.

Proof. Let N be a submodule of M equipped with the induced filtration $F_pN = F_pM \cap N$. Since $G(N) \subset G(M)$, $G(N)$ is finitely generated. Proposition 5.3. entails that N is finitely generated, so M is left Noetherian. Consider submodules N and P of M, $N \subset P$, both equipped with the induced filtration, and assume that $G(N) = G(P)$. If $\{x_{p_i}^{(i)}, 1 \leq i \leq n\}$ is a set of homogeneous generators for $G(N)$, where $x^{(i)} \in N$ for all i, $1 \leq i \leq n$, then, again from Proposition 5.3., we infer that $\{x^{(i)}, 1 \leq i \leq n\}$ generates N as well as P. Thus $N = P$. Lemma 5.4 applies and yields $\text{K-dim}_R M \leq \text{K-dim}_{G(R)} G(M)$. In order to establish our last claim, let $N \neq o$, $N \subset M$ and filter M/N by putting : $F_p(M/N) = (N + F_pM)/N$. Clearly, we obtain an exact sequence in G(R)-gr

$$o \longrightarrow G(N) \longrightarrow G(M) \longrightarrow G(M/N) \longrightarrow o \quad .$$

Since $G(N) \neq o$, the Krull dimension of $G(M/N)$ is less than α, hence :

$$\text{K-dim}_R M/N \leq \text{K-dim}_{G(R)} G(M/N) < \alpha \quad .$$

This shows exactly that M is an α-critical module. □

5.7. Corollary. Let R be a complete filtered ring such that G(R) is a left Noetherian ring, then :

1°. R is left Noetherian and K-dim $R \leq$ K-dim $G(R)$

2°. F_oR is left Noetherian and K-dim $F_oR \leq$ K-dim $G(R) + 1$.

Proof. 1. follows from the foregoing corollary.

2. F_oR is a subring of R and it is clearly complete; Corollary 5.6. together with Proposition 3.2 and Corollary 4.12 of Chapter II will finish the proof. □

5.8. Proposition. Let $M \in$ R-filt with FM being exhaustive. Assume that submodules of M are closed, i.e. if $N \subset M$ then $N = \bigcap_{p \in \mathbb{Z}} (N + F_pM)$. Consider submodules $N \subset P \subset M$, then :

1. If $G(N) = G(P)$ then $N = P$. In particular if the Krull dimension of G(M) is well defined then the Krull dimension of M is defined and $\text{K-dim}_R M \leq \text{K-dim}_{G(R)} G(M)$.

2. If $\text{K-dim}_R M = \text{K-dim}_{G(R)} G(M) = \alpha$ and if G(M) is α-critical then M is an α-critical

module.

3. If $G(M)$ may be generated by n homogeneous generators then M may be generated by n generators.

Proof. 1. Take $x \in P$, then $x \in F_iM$ for some i, $x \notin F_{i-1}M$. Because : $x_i \in G(P)_i = G(N)_i = (P \cap F_iM) + F_{i-1}M/F_{i-1}M = (N \cap F_iM) + F_{i-1}M/F_{i-1}M$, there is a $y_1 \in N \cap F_iM$ such that $x - y_1 \in F_{i-1}M$. Hence $x \in y_1 + P \cap F_{i-1}M$. Repeating this proces we end up with elements $y_1, y_2, \ldots, y_0 \in N$ such that $x - (y_1 + y_2 + \ldots + y_s)$ is in $F_{i-s}M \cap P$, $s > o$. Thus $x \in N + F_{i-s}M$ for $s > o$. It follows that $x \in \bigcap_{p \in \mathbb{Z}} (N + F_pM) = N$, whence $N = P$ follows.

2. Proceed as in the proof of Corollary 5.4.

3. Let $\{x^{(i)}_{p_i}, 1 \leq i \leq n\}$ be a family of homogeneous generators for $G(M)$. Write M' for the submodule of M generated by the elements $x^{(i)}, 1 \leq i \leq n$. Obviously, $x^{(i)}_{p_i} \in G(M')_{p_i} = (M' \cap F_{p_i}M)/(M' \cap F_{p_i-1}M)$. Therefore $G(M') = G(M)$ and by 1. this yields $M' = M$.

5.9. Remark. The above proposition is generally applied in case FM is exhaustive and discrete.

Let R be a ring, I an ideal of R, $M \in R$-mod. Using the I-adic filtration, as introduced in 1.2.2°, we get the I-adic completion $\hat{M}$ of M by $\hat{M} = \varprojlim_n M/I^nM$. Let $i : M \to \hat{M}$ be the canonical filtered morphism. In studying I-adic completion the following property is very relevant. I is said to satisfy the left Artin-Rees property if for any finitely generated left R-module M, any submodule N of M and any natural number n, there exists an integer $h(n) \geq o$ such that $I^{h(n)}M \cap N \subset I^nN$. In other words, I satisfies the left Artin-Rees property if the I-adic topology of N coincides with the topology induced in N by the I-adic topology of M.

5.10. Proposition. Let R be a left Noetherian ring and I an ideal satisfying the left Artin-Rees property.

1. The functor carrying M to $\hat{M}$ is exact in the category of left R-modules of finite type.
2. $\hat{R}$ is a right flat R-module.
3. If $M \in R$-mod is finitely generated then $\hat{M} = \hat{R} \otimes_R M$.

4. If $M \in R$-mod is finitely generated then $\operatorname{Ker} i = \bigcap_{n \geq 1} I^n M = \{x \in M,\ (1-r)x = 0$ for some $r \in I\}$.

   In particular if I is contained in the Jacobson radical of R then $\bigcap_{n \geq 1} I^n M = 0$.
5. If $M \in R$-mod is finitely generated then $\hat{M} = \hat{R}.i(M)$ and $\hat{M}$ is a finitely generated $\hat{R}$-module.

Proof. cf. [3], chapter 3, Theorem 3 p.68; Proposition 5 p.65. □

5.11. Corollary. If R is a left Noetherian ring, I an ideal of R satisfying the left Artin-Rees property, then the following statements are equivalent :

1. I is contained in the Jacobson radical of R.
2. If $M \in R$-mod is finitely generated then submodules of M are closed in the I-adic topology.

An ideal I of R is said to be generated by a central system if $I = Rx_1 + \dots + R_{x_n}$ where $x_1 \in Z(R)$ and where the image of x_i in $R/(x_1,\dots,x_{i-1})$ is in the center of the latter ring.

5.12. Proposition. Let R be a left Noetherian ring, I an ideal generated by a central system, then I satisfies the left Artin-Rees property.

Proof. Let N be a submodule of a finitely generated $M \in R$-mod. Let $s \in \mathbb{N}$ and consider a submodule M' of M maximal with the property that $M' \cap N = I^s N$. This choice makes M/M' into an essential extension of $N/I^s N$, where $I^s\ (N/I^s N) = 0$. If we are able to establish that under these circumstances M/M' may be annihilated by some I^t then $I^t(M/M') = 0$ yields $I^t M \cap N \subset I^s N$. Note also that it is sufficient to do this for $s = 1$ because for arbitrary s it will follow from this case applied to $IN, \dots I^{s-1}N$ instead of N and combination of the inclusions thus obtained. If $I = Rc_1 + \dots + Rc_n$, where $(c_1,\dots,c_n)$ is a central system, let $\mu : M \to M$ be the map given by $m \to c_1 m$ for all $m \in M$. Since M is left Noetherian, there exists $r \in \mathbb{N}$ such that $\operatorname{Ker} \mu^r = \operatorname{Ker} \mu^{r+k}$ for all $k \in \mathbb{N}$ i.e. $\operatorname{Im} \mu^r \cap \operatorname{Ker} \mu^r = 0$. However N is contained in $\operatorname{Ker} \mu^r$ and N is an essential submodule of M, therefore $\operatorname{Im} \mu^r = 0$. Let t be the smallest natural number such that $\mu^t = 0$, then μ gives rise to an injective R-morphism $M/\operatorname{Ker} \mu \to \operatorname{Ker} \mu^{t-1}$. To prove our original claim it will therefore be sufficient to prove that both Ker μ and Ker μ^{t-1} can be annihilated by large powers of I. Now c_1 being central, Ker μ is an R/Rc_1-module and we apply induction on n to deduce that I^p Ker $\mu = 0$ for some $p \in \mathbb{N}$. Since there is an injective morphism : Ker μ^m / Ker μ $\rightarrow$ Ker μ^{m-1} for $1 \leq m \leq t$, the fact that I^q Ker $\mu^{t-1} = 0$ for some $q \in \mathbb{N}$ will follow by induction on t. □

5.13. Example. If $\mathfrak{g}$ is a nilpotent Lie algebra, finite dimensional over a field of characteristic zero, then any ideal of the enveloping algebra $U(\mathfrak{g})$ is generated by a central system.

$P \in$ R-filt is said to be filt-projective if, the diagram in R-filt where f is a strict morphism

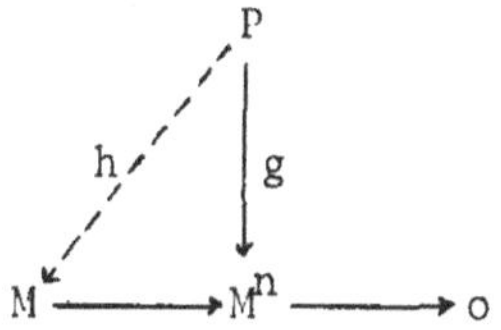

may be completed by a filtered morphism h such that the diagram is commutative o

5.14. Proposition. Let R be a filtered ring, $P \in$ R-filt such that FP is exhaustive, then the following statements are true :

1. P is filt-projective if and only if P is a direct summand in R-filt of a filt-free object.
2. If P is filt-projective, then G(P) is projective in G(R)-gr.
3. If FR is complete and if P is finitely generated filt-projective then FP is complete.

Proof. 1. Let L be filt-free with filt-basis $(x^{(i)}, n_i)_{i \in J}$. Consider the following diagram in R-filt :

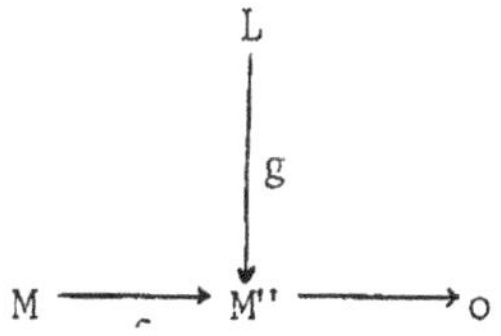

where f is a strict epimorphism. Since $x^{(i)} \in F_{n_i}L$, it follows that $g(x^{(i)}) \in F_{n_i}M'' = f(F_{n_i}M)$. Hence there exists $y^{(i)} \in F_{n_i}M$ such that $g(x^{(i)}) = f(y^{(i)})$. Lemma 5.1.5° yields the existence of a filtered morphism of degree o, $h : L \to M$ such that $h(x^{(i)}) = y^{(i)}$. Clearly $g = f \circ h$, hence L is filt-projective. If P is any filt-projective object then there exists a filt-free object L and a strict epimorphism $L \xrightarrow{f} P \to o$ (see Lemma 5.1.8°). Projectivity of P implies the existence of a morphism $g' : P \to L$ of degree o, with $f \circ g' = 1_P$. We have $g'(F_iP) \subset F_iL \cap \mathrm{Im}\, g'$ while conversely $x \in F_iL \cap \mathrm{Im}\, g'$ implies $x.g'(y) \in F_iL$ i.e. $f(x) = f \circ g'(y) = y \in F_iP$. Therefore $x \in g'(F_iP)$ or $g'(F_iP) = F_iL \cap \mathrm{Im}\, g'$ follows and the latter means that g' is strict. Note first that in R-mod we have : $L = \mathrm{Im}\, g' \oplus \mathrm{Ker}\, f$. Now we claim that :

$$F_iL = F_i\,\mathrm{Im}\,g' \oplus F_i\,\mathrm{Ker}\,f = (\mathrm{Im}\,g' \cap F_iL) \oplus (\mathrm{Ker}\,f \cap F_iL).$$

Indeed, if $x \in F_iL$ then $x = y + z$ with $y \in \mathrm{Im}\, g'$ and $z \in \mathrm{Ker}\, f$. Hence $f(y) = f(x) \in F_iP$ and, while $y \in \mathrm{Im}\, g'$, we get that $g' \circ f(y) = y$ or $y \in g'(F_iP) = F_iL \cap \mathrm{Im}\, g'$. On the other hand, $z = x - y \in F_iL \cap \mathrm{Ker}\, f$ and thus $L = \mathrm{Im}\, g \oplus \mathrm{Ker}\, f$ in R-filt. Finally since g' is a strict monomorphism and since we have $\mathrm{Im}\, g' \cong P$ in R-filt it follows that P is a direct summand of a filt-free object.

The converse implication is easy.

2. Directly from 1 and the properties of the functor G.

3. The construction of L in 1 shows that, in case P is finitely generated in R-filt, we may choose L to be finitely generated too and we have that $L = P \oplus Q$ for some $Q \in$ R-filt. Since R is complete, L is complete and so a Cauchy-sequence $(x_n)_{n \in \mathbb{N}}$ in P converges to some $x \in L$. Write $x = p + q$ with $p \in P$, $q \in Q$. Since FP is exhaustive, $p \in F_kP$ for some $k \in \mathbb{Z}$ and it is clear that $x_n - p$ will be in F_tL for t large enough i.e. $q = \lim\limits_{\overrightarrow{n}} (x_n - p) = o$ and $x \in P$ or P is complete follows. □

I.6. The Functor $HOM_R(-,-)$.

The properties of HOM_R mentioned in I.2 make it into a functor R-filt x R-filt $\to \mathbb{Z}$-filt; and it is clear that, by definition of the filtration in $HOM_R(-,-)$,

$F\,HOM_R(M,N)$ is exhaustive for any $M,N \in R$-filt. First let us claim the analogies of Lemma 3.3.2 and Remark 3.3.3.

<u>6.1. Lemma.</u> Let R be a filtered ring and let $M,N \in R$-filt be such that M is a finitely generated object while FN is exhaustive, then : $HOM_R(M,N) = Hom_R(M,N)$.

<u>Proof.</u> Let $\{x^{(i)}, n_i\}$ be a system of generators for M and take $f \in Hom_R(M,N)$. There exists $s \in \mathbb{Z}$ such that $f(x^{(i)}) \in F_{n_i+s}N$ for any $1 \leq i \leq n$. It easily derives from this that $f \in F_s\,HOM_R(M,N)$ and hence $f \in HOM_R(M,N)$. Note that, if FN is not exhaustive and if N' is the exhaustion of N then $HOM_R(M,N) = HOM_R(M,N') = Hom_R(M,N')$.

<u>6.2. Remark.</u> In general $HOM_R(M,N) \neq Hom_R(M,N)$. For example let R be a discretely and exhaustively filtered ring such that $R = F_iR$ for all i when $F_i(R) \neq 0$. Take M to be filt-free with filt-basis $\{x^{(i)}, o\}$ with $1 \leq i \leq \infty$, and take $N = R$. Define $f : M \to N$ by $f(x^{(i)}) = a^{(i)}$ where $a^{(i)} \notin F_iR$. If the degree of f is p, let $i_o \in \mathbb{Z}$ be such that $F_iN = o$ for any $i < i_o$. Consider $i < i_o - p$, hence $f(x^{(i)}) = a^{(i)} = o$, contradiction.

<u>6.3. Proposition.</u> Let $M,N \in R$-filt, then :

1. If FM is exhaustive and FN is separated, then $FHOM_R(M,N)$ is separated.
2. If M is a finitely generated R-module, if FM is exhaustive and FN is discrete then $FHOM_R(M,N)$ is discrete.
3. If FM is exhaustive and FN is complete then $FHOM_R(M,N)$ is complete.

<u>Proof.</u> 1. Let $f \in \bigcap_p F_p HOM_R(M,N)$. Then $f(F_pM) \subset \bigcap_s F_{p+s}N = o$. Therefore $f(M) = f^p(\bigcup_p F_pM) = o$.

2. Let $x^{(1)}, \ldots, x^{(n)}$ be generators for M. Since FM is exhaustive, there is an $r_o \in \mathbb{Z}$ such that $x^{(i)} \in F_{r_o}M$ for all $i \in \{1, \ldots, n\}$. Since $F_iN = o$ for all $i < n_o$ for some $n_o \in \mathbb{Z}$, it follows that for any $f \in F_iHOM_R(M,N)$ where $i < n_o - r_o$ we have $f(x^{(i)}) = o$, hence $F_iHOM_R(M,N) = o$ if $i < n_o - r_o$.

3. Let $q < p \in \mathbb{Z}$ and consider the projective system :

$$\begin{array}{ccc} & HOM_R(M,N) & \\ {\scriptstyle\varphi_p}\swarrow & & \searrow{\scriptstyle\varphi_q} \\ HOM_R(M,N)/F_{-p}HOM_R(M,N) & \xrightarrow{\ \pi_{pq}\ } & HOM_R(M,N)/F_{-q}HOM_R(M,N) \end{array} .$$

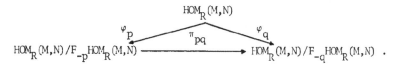

We claim that $HOM_R(M,N) = \varprojlim HOM_R(M/N)/F_{-p}HOM_R(M,N) = X$. Indeed, let $(f^{(p)}, p\in\mathbb{Z})$ be an element of X such that $^p\pi_{pq}(f^{(p)}) = f^{(q)}$ for $p>q$. Let $f^{(p)} = \varphi_p(g^{(p)})$, with $g^{(p)} \in HOM_R(M,N)$. If $q<p$ then $g^{(p)} - g^{(q)} \in F_{-q}HOM_R(M,N)$ and therefore $(g^{(p)} - g^{(q)})(F_sM) \subset F_{s-q}N$ for any $s\in\mathbb{Z}$. Since FM is exhaustive, for any $x\in M$ there exists some $t\in\mathbb{Z}$ such that $x\in F_tM$ and thus $g^{(p)}(x) - g^{(q)}(x) \in F_{t-q}N$, consequently the sequence $\{g^{(p)}(x), p\in\mathbb{Z}\}$ is a Cauchy sequence in N. Completeness of N entails that the mapping $f : M\to N$ defined by $f(x) = \lim_{p\to\infty} g^{(p)}x$, is well defined. If $x\in F_tM$, knowing that $g^{(p)}(x) \in g^{(q)}(x) + F_{t-q}N$ for $p\geq q$, we may conclude that $f(x) \in g^{(q)}(x) + F_{t-q}N$. Therefore $(f - g^{(q)})(F_tM) \subset F_{t-q}N$ or $f - g^{(q)} \in F_{-q}HOM_R(M,N)$. Hence $f\in HOM_R(M,N)$ and $\varphi_q(f) = f^{(q)}$ for the selected q. □

If $M,N\in R\text{-filt}$, we introduce a natural map $\varphi = \varphi(M,N)$,

$$\varphi : G(HOM_R(M,N)) \longrightarrow HOM_{G(R)}(G(M),G(N)) ,$$

as follows : for $f\in F_pHOM_R(M,N)$, $x\in F_qM$ put $\varphi(f_p)(x_q) = f(x)_{p+q}$.

6.4. Lemma. $\varphi(M,N)$ is a monomorphism. Moreover, $\varphi(M,N)$ is an isomorphism if M is filt-projective.

Proof. Fixing $M,N\in R\text{-filt}$ we write φ instead of $\varphi(M,N)$. If $\varphi(f_p) = o$, then $f(x)_{p+q} = o$ for every $x\in F_qM$, $q\in\mathbb{Z}$. It follows that $f(F_qM) \subset F_{p+q-1}N$ for any $q\in\mathbb{Z}$ i.e. $f\in F_{p-s}HOM_R(M,N)$. Consequently $f_p = o$. In order to prove the second statement, first assume that M is filt-free and let $(x^{(i)}, p_i)_{i\in J}$ be a filt-basis for M. In this case $\{x^{(i)}_{p_i}, i\in J\}$ is a homogeneous basis for G(M). Letting $g\in HOM_{G(R)}(G(M),G(N))_p$, we obtain $g(x^{(i)}_{p_i}) = y^{(i)}_{p+p_i}$. Define $f : M\to N$ by putting $f(x^{(i)}) = y^{(i)}$; with this definition it is clear that $f\in F_pHOM_R(M,N)$ and also that $\varphi(f_p) = g$, proving surjectivity of φ. If M is filt-projective then there is a filt-free L such that $L = M\oplus M'$ in R-filt. We know that HOM_R commutes with finite direct sums, therefore we may use the previous part of the proof. □

6.5. Definition. Let FR be an exhaustive filtration of R. A filt-injective object is an object $M\in R\text{-filt}$ such that, if I is a left ideal of R equipped with the induced filtration then any filtered morphism $f : I \to M$ extends to a filtered morphism $R \to M$, equivalently, the canonical mapping $HOM(i, 1_M) : HOM_R(R,M) \to HOM_R(I,M)$ is an epimorphism.

6.6. Remarks. 1. Let FR be exhaustive and complete and suppose that $G(R)$ is left Noetherian. If M is R-filt injective then it is injective in R-mod, because of Lemma 6.1.

2. Let FR be exhaustive. If $Q \in R$-filt is injective in R-mod then it is injective in R-filt.

6.7. Theorem. Let FR be exhaustive and let $Q \in R$-filt be complete. If $G(Q)$ is injective in $G(R)$-gr then Q is filt-injective.

Proof. Equip I, left ideal of R, with the filtration $F_pI = F_pR \cap I$. Since $i : I \to R$ is a strict morphism we obtain the following commutative diagram :

$$\begin{array}{ccc} G(HOM_R(R,Q)) & \xrightarrow{G\, HOM_R(i,1_Q)} & G(HOM_R(I,Q)) \\ \Big\downarrow \varphi(R,Q) & & \Big\downarrow \varphi(I,Q) \\ HOM_{G(R)}(G(R),G(Q)) & \xrightarrow{HOM_{G(R)}(G(i),G(1_Q))} & HOM_{G(R)}(G(I),G(Q)) \end{array} .$$

Since $G(i)$ is a monomorphism, injectivity of $G(Q)$ entails that $HOM_{G(R)}(G(i),G(I_Q))$ is an epimorphism. Hence $\varphi(I,Q)$ is an epimorphism too, therefore an isomorphism. Thus $G\,HOM_R(i,1_Q)$ is an epimorphism, while $F\,HOM_R(R,Q)$ and $F\,HOM_R(I,Q)$ are exhaustive and complete, therefore by Theorem 4.4.3., $HOM_R(i,1_Q)$ is epimorphic. □

I.7. Projective Modules and Homological Dimension of Rings.

We will say that $x \in R$ is topologically nilpotent if the sequence $\{x^n, n \in \mathbb{N}\}$ is a Cauchy sequence converging to o. Recall the following lemma about topological rings :

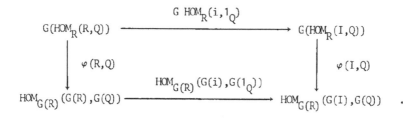

7.1. Lemma. Let R be a complete topological ring and consider a fundamental set of neighborhoods of zero, consisting of additive subgroups. If $\varphi : R \to S$ is a ring homomorphism such that every $x \in \operatorname{Ker}\varphi$ is topologically nilpotent, then every idempotent element of $\operatorname{Im}\varphi$ may be lifted to R.

Proof. Let $e \in \operatorname{Im}\varphi$ be idempotent, $e = \varphi(x)$ for some $x \in R$. Since $\varphi(x^2 - x) = 0$, $a = x^2 - x \in \operatorname{Ker}\varphi$. Put $b = x + \lambda(1-2x)$ with $\lambda \in R$, and determine λ such that $b^2 = b$ while λ commutes with a. We get : $(\lambda^2 - \lambda)(1 + 4a) + a = 0$.

Put : (*) $\lambda = \frac{1}{2}(1 - (1+4a)^{\frac{1}{2}}) = \frac{1}{2} \sum_{1 \leq k < \infty} (-1)^{k-1} c_{2k}^{k} a^{k}$, where c_{2k}^{k} is the coefficient of $1^k(-1)^k$ in the binomial expansion of $(1-1)^{2k} = 0$ i.e. $\frac{1}{2} c_{2k}^{k}$ is an integer. Therefore λ may be viewed as a power series with integer coefficients which is easily seen to converge. Hence (*) determines λ and it commutes with a and consequently b is idempotent. Furthermore, $a \in \operatorname{Ker}\varphi$ yields $\lambda \in \operatorname{Ker}\varphi$ and thus $\varphi(b) = \varphi(x) = e$. □

7.2. Lemma. Let R be a filtered ring, $M \in R$-filt such that FM is exhaustive and complete. Suppose that $f : M \to M$ is a filtered morphism such that $G(f)^2 = G(f)$. Then there exists a strict morphism $g : M \to M$ in R-filt satisfying : $g^2 = g$ and $G(g) = G(f)$. The proof follows from the foregoing plus the fact that every idempotent f in $\mathrm{HOM}_R(M,M)$ is a strict morphism. Indeed, let $f(x) \in F_pM \cap \operatorname{Im} f$, then $f(f(x)) = f(x)$ yields $f(x) \in f(F_pM)$ hence f is strict. □

7.3. Remark. If $M \in R$-filt is such that FM is exhaustive and complete then $F_{-1}\mathrm{HOM}_R(M,M) \subset J(\mathrm{HOM}_R(M,M))$ (J = Jacobson radical). Indeed if $f \in F_{-1}\mathrm{HOM}_R(M,M)$ then f is topologically nilpotent and so 1-f has an inverse $\sum_{0 \leq n < \infty} f^n$.

7.4. Corollary. If M is as in the foregoing lemma then any countable set of orthogonal idempotent elements in $\operatorname{Im}\varphi$ can be lifted to $\mathrm{Hom}_{R\text{-filt}}(M,M)$.

7.5. Theorem. Let R be a filtered ring such that FR is exhaustive. Let P_g be projective in G(R)-gr and suppose that, either FR is discrete and FP is left-limited or P is finitely generated while FR is complete. Then there is a filt-projective module $P \in R$-filt such that $G(P) = P_g$. If $M \in R$-filt, then for any morphism $g : P_g \to G(M)$ of degree p there is a filtered morphism $f : P \to M$ of degree p such that $g = G(f)$.

Proof. The hypotheses imply that there is a free object L_g in $G(R)$-gr and a morphism $h : L_g \to L_g$ of degree zero, such that $h^2 = h$ and $\mathrm{Im}\, h = P_g$. If P_g is finitely generated then L_g may be chosen to be finitely generated too. Assume that $L_g = G(L)$, where $L \in R$-filt is filt-free. If P_g has left limited grading then L_g may be chosen so that it has left-limited grading too and we may suppose FL to be discrete since FR is discrete. In the case where P_g is finitely generated and FR is complete, we obtain that FL is complete. By Lemma 7.2, there exists a strict filtered morphism $f : L \to L$ with $f^2 = f$ and $G(f) = h$. Put $P = \mathrm{Im}\, f$. Since f is strict it follows that $G(P) = P_g$ and that P is filt-projective as required. Moreover if P is finitely generated then FP is complete. To prove the remaining statement we may write $g : G(P) \to G(M)$ and up to taking a suitable suspension we may assume $\deg g = o$. We have $L = P \oplus Q$ and hence there is a graded morphism $h : G(L) \to G(M)$ such that $h|G(P) = g$. However $h = G(k)$ for some filtered morphism $k : L \to M$. Putting $f = h|P$ we arrive at $g = G(f)$. □

7.6. Corollary. Let R be a filtered ring such that FR is exhaustive and discrete and let $M \in R$-filt be such that FM too is exhaustive and discrete. Then we have :

$$\mathrm{p.dim}_R M \leqslant \mathrm{p.dim}_{G(R)}\, G(M) \; .$$

Proof. By Lemma 5.1.8°, there exists a free resolution for M in R-filt :

$$\cdots \longrightarrow L_{n-1} \xrightarrow{f_{n-1}} L_{n-2} \longrightarrow \cdots \longrightarrow L_1 \xrightarrow{f_1} L_o \xrightarrow{f_o} M \longrightarrow o$$

where FL_j are discrete for all j. Write Q for Ker f_{n-1}, where $n = \mathrm{p.dim}_{G(R)} G(M)$, then $G(Q)$ is a projective $G(R)$-module with left-limited grading. By the theorem we may select a projective object $Q \in R$-filt such that $G(Q) \cong G(Q')$, let $g : G(Q') \to G(Q)$ denote this isomorphism. There exists a filtered morphism $f : Q' \to Q$ such that $G(f) = g$. Now we end up with the situation where Q' is filt-projective, FR is discrete (note that FQ' turns out to be discrete just as well) while G(f) is an isomorphism, therefore we may apply Corollary 4.5 and deduce that f is an isomorphism, whence it follows that Q is filt-projective and therefore projective in R-mod. □

7.7. Corollary. Let R be a filtered ring with exhaustive and discrete filtration, then gl.dim $R \leq$ gr.gl.dim $G(R)$.

Proof. Equip a left R-module M with the filtration $F_iM = F_iR.M$, then FM is exhaustive and discrete. By the foregoing corollary we have : p.dim $M \leq$ p.dim $G(M) \leq$ gr.gl.dim $G(R)$. □

7.8. Corollary. Let R be a graded ring, then : gr.gl.dim $R \leq$ gl.dim $\underline{R}$. If the gradation of R is left limited then : gr.gl.dim $R =$ gl.dim $\underline{R}$.

Proof. The second statement follows from the foregoing corollary applied to the ring R filtered with the associated filtration. The first statement follows directly from Corollary 3.3.12. □

7.9. Corollary. Let R be a filtered ring such that FR is exhaustive and complete, let $M \in R$-filt be such that FM is exhaustive. Suppose that G(R) and G(M) are left Noetherian, then: $\text{p.dim}_R M \leq \text{p.dim}_{G(R)} G(M)$.

Proof. Apply Lemma 5.1 and Theorem 7.5 and use the argumentation of the proof of Corollary 7.6. □

7.10. Corollary. Let R be a filtered ring such that FR is exhaustive and complete. Suppose that G(R) is left Noetherian. We have : gl.dim $R \leq$ gr.gl.dim $G(R)$.

Proof. Let M be a finitely generated left R-module and take $x_1, \ldots, x_m$ to be a set of generators. Filter M with the filtration : $F_iM = \sum_{k=1}^{n} F_iR.x_k$. In doing this we obtain a finitely generated filtered module with an exhaustive filtration. Corollary 7.8 finishes the proof. □

I.8. Weak (flat) Dimension of Filtered Modules.

Let us start this section with a basic lemma on tensor products of filtered modules. If $M \in R$-filt, $N \in$ filt-R then the $\mathbb{Z}$-module $N \otimes_R M$ has filtration : $F_p(N \otimes_R M)$ is the $\mathbb{Z}$-submodule of $N \otimes_R M$ generated by all elements of type $n \otimes m$ where $n \in F_tN$,

$m \in F_s M$ and $p = s+t$. This construction yields a functor :

$$\underset{R}{\otimes} : \text{filt-}R \times R\text{-filt} \to \mathbb{Z}\text{-filt} .$$

8.1. Lemma. With notations as above, the following statements are true :

1°. If FM and FN are exhaustive then so is $F(N \underset{R}{\otimes} M)$.

2°. If FM and FN are discrete then so is $F(N \underset{R}{\otimes} M)$.

3°. For all $m,n \in \mathbb{Z}$: $N(n) \underset{R}{\otimes} M(m) = (N \underset{R}{\otimes} M)(m+n)$.

4°. In the category R-filt we have the following isomorphism $R \underset{R}{\otimes} M \cong M$; similarly $N \otimes_R R \cong N$ in filt-R.

5°. The functor $\underset{R}{\otimes}$ commutes with direct sums and inductive limits.

6°. For $M \in$ filt-R, $N \in R$-filt-S, $P \in$ filt-S, we have the following isomorphism in $\mathbb{Z}$-filt : $HOM_S(M \otimes_R N, P) \cong HOM_R(M, HOM_S(N,P))$.

Proof. Statements 1°,2°,3° are clear.

4°. Consider the R-module morphism, $\varphi : N \underset{R}{\otimes} R \;\; N$, $\varphi(x \otimes r) = xr$. Its inverse is given by, $\psi : N \to N \otimes_R R$, $\psi(x) = x \otimes 1$. Now if $x \otimes r \in F_p(N \otimes_R R)$ with $x \in F_i M$, $r \in F_j R$, $i+j = p$, then $\varphi(x \otimes r) = xr \in F_i N . F_j R \subset F_p N$. In a similar way one establishes that ψ is a morphism in filt-R.

5°. Let $N = \bigoplus_{\alpha \in A} N_\alpha$, $M = \bigoplus_{\beta \in B} M_\beta$, then $\bigoplus_{(\alpha,\beta) \in A \times B} N_\alpha \otimes_R M \cong N_R \otimes M$. Indeed, let

$$\varphi : N \otimes_R M \longrightarrow \bigoplus_{(\alpha,\beta) \in A \times B} (N_\alpha \otimes_R M_\beta)$$

be the R-module morphism which is given by :

$$\varphi\left(\left(\sum_{\alpha \in A} x_\alpha\right) \otimes \left(\sum_{\beta \in B} y_\beta\right)\right) = \sum_{(\alpha,\beta) \in A \times B} x_\alpha \otimes y_\beta .$$

Then it is well known that φ is an isomorphism of abelian groups. Let $x \otimes y \in F_p(N \otimes_R M)$ where $x \in F_i N$, $y \in F_j N$ and $i+j = p$. If $x = \sum_{\alpha \in A} x_\alpha$, $y = \sum_{\beta \in B} y_\beta$ then $x_\alpha \in F_i N_\alpha$ and $y_\beta \in F_j M_\beta$. It follows immediately that $x_\alpha \otimes y_\beta \in F_p(N_\alpha \otimes_R M_\beta)$, hence $\varphi(x \otimes y) \in F_p(\underset{(\alpha,\beta) \in A \times B}{\bigcirc} N_\alpha \otimes M_\beta)$. In a similar way one establishes that φ^{-1} is actually a morphism in $\mathbb{Z}$-filt. The assertion for inductive limits may be proved in an equally straightforward way.

6°. Define

$$\varphi : \mathrm{HOM}_S(M \otimes_R N, P) \longrightarrow \mathrm{HOM}_R(M, \mathrm{HOM}_S(N,P))$$

by : $(\varphi(f)(x))(y) = f(x \otimes y)$, and define

$$\psi : \mathrm{HOM}_R(M, \mathrm{HOM}_S(N,P)) \longrightarrow \mathrm{HOM}_S(M \otimes_R N, P)$$

by : $\psi(g)(x \otimes y) = (g(x))(y)$. If $f \in F_p\mathrm{HOM}_S(M \otimes_R N, P)$, $x \in F_t M$, $y \in F_\sigma N$ then it follows from $x \otimes y \in F_{t+r}\, M \otimes_R N$ that $f(x \otimes y) \in F_{p+t+r}P$. Hence $\varphi(f)(x) \in F_{p+t}\, \mathrm{HOM}_S(N,P)$ and thus $\varphi(f) \in F_p\mathrm{HOM}_R(M, \mathrm{HOM}_S(N,P))$. This expresses that φ is a morphism in $\mathbb{Z}$-filt. Along the same lines one verifies that ψ is a morphism of degree o and ψ is the inverse of φ. □

Let $N \in R$-filt, $M \in$ filt-R. Define a graded morphism

$$\varphi = \varphi(N,M) : G(M) \underset{G(R)}{\otimes} G(N) \longrightarrow G(M \otimes N),$$

by putting : $\varphi(m_s \otimes n_t) = (m \otimes n)_{s+t}$. Obviously this is well defined and it is clear that φ is surjective

<u>8.2. Lemma.</u> If either M or N is a free object in filt-R or R-filt respectively then $\varphi(N,M)$ is an isomorphism.

<u>Proof.</u> Knowing that $\otimes_R$ and G commute with direct sums, the proof may be reduced to the case where $M = R(n)_R$. But then Lemma 8.1 allows us to reduce the proof further to the case $M = R_R$. In this case we have the following commutative diagram of graded morphisms, vertical arrows representing isomorphisms.:

$$\begin{array}{ccc} G(R_R) \underset{G(R)}{\otimes} G(N) & \xrightarrow{\varphi(M,R)} & G(R \otimes_R N) \\ \uparrow & & \uparrow \\ G(N) & \xrightarrow{G(N)} & G(N) \end{array}$$

So, $\varphi(M,R)$ is also an isomorphism. □

8.3. Lemma. Let R be a filtered ring and let $M \in R$-filt be such that FR and FM are discrete and exhaustive. If G(M) is a flat object in G(R)-gr then M is a flat R-module.

Proof. If J is a right ideal of R, equip it with the filtration $F_pJ = F_pR \cap J$. Let $i : J \to R$ be the canonical inclusion morphism; note that i is a strict morphism. Application of Theorem 4.4 yields that G(i) is injective, and so we obtain the following commutative diagram :

$$\begin{array}{ccc} G(J) \underset{G(R)}{\otimes} G(M) & \xrightarrow{G(i)\otimes 1_{G(M)}} & G(R) \underset{G(R)}{\otimes} G(M) \\ \downarrow \varphi(M,J) & & \downarrow \varphi(M,R) \\ G(J \otimes_R M) & \xrightarrow[G(i\otimes 1_M)]{} & G(R \otimes_R M) \end{array}$$

The fact that $\varphi(M,R)$ is an isomorphism, whereas $G(i)\otimes 1_{G(M)}$ is a monomorphism, entails that $\varphi(M,J)$ is a monomorphism, hence an isomorphism. Then $G(i\otimes 1_M)$ has to be monomorphic and repeated application of Theorem 4.4. allows to derive from this that $F(J\otimes_R M)$ is discrete. Hence, $i\otimes 1_M$ is a strict monomorphism and it follows that M is flat. □

8.4. Corollary. Let R be a filtered ring, $M \in R$-filt, such that FR and FM are discrete and exhaustive. Then we have :

$$\text{w.dim}_R M \leq \text{gr.w.dim}_{G(R)} G(M).$$

Proof. Put $n = \text{gr.w.dim}_{G(R)} G(M)$. Consider an exact sequence :

$$o \longrightarrow L \xrightarrow{f} L_{n-1} \longrightarrow \dots \longrightarrow L_1 \xrightarrow{f_1} L_o \xrightarrow{f_o} M \longrightarrow o ,$$

where $L_o, \dots, L_{n-1}$ are free objects, while FL_i, $i = o, \dots, n-1$, and FL are exhaustive and discrete. By exactness of the functor G we derive from (*) an exact sequence :

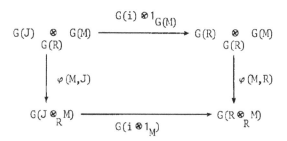

$$o \longrightarrow G(L) \longrightarrow G(L_{n-1}) \longrightarrow \dots \longrightarrow G(L_o) \longrightarrow G(M) \longrightarrow o \quad .$$

Our assumptions imply that G(L) is a flat object of G(R)-gr and the foregoing lemma entails that L is flat. Therefore $w.\dim_R M \leq n$. □

8.5. Corollary. For a filtered ring R with FR being exhaustive and discrete we have :

$$gl.w.\dim R \leq gr.gl.w.\dim G(R) \ .$$

Proof. Put $n = gr.gl.w.\dim G(R)$. Any $M \in$ R-mod may be filtered by the filtration $F_pM = F_pR.M$ which is easily seen to be exhaustive and discrete. The inequality of Corollary 8.4. thus holds for any M, filtered this way, an this proves our claim. □

8.6. Corollary. Let R be a graded ring, then :

$$gr.gl.w.\dim.R \leq gl.w.\dim \underline{R}$$

If moreover R is left limited then :

$$gr.gl.w.\dim R = gl.w.\dim \underline{R} \ .$$

Proof. Easy combination of foregoing results.

8.7. Corollary. Let R be a positively graded ring and assume that R_o is a von Neumann regular ring : $gl.w.\dim R = \text{left-}w.\dim_R R_o = \text{right-}w.\dim_R R_o$.

Proof. Take $M \in$ R-mod and equip it with the trivial filtration $F_pM = o$ for $p < o$, $F_pM = M$ for $p \geq o$. Endow R with the filtration $F_pR = \underset{i \leq p}{\otimes} R_i$. In this setting M is a filtered R-module and clearly $G(M)_p = o$ for all $p \neq o$. Therefore G(M) is annihilated by the ideal $R_+ = \underset{n \geq 1}{\oplus} R_n$. Corrolary 8.4 together with [2 ,ex.5 p.360] we get : $w.\dim_R M \leq w.\dim_R G(M) \leq 1.w.\dim_R R_o + 1.w.\dim_{R_o} G(M)$ and the latter is just $1.w.\dim_R R_o$. It follows that $gl.w.\dim R = 1.w.\dim_R R_o$. The other equality may be established in exactly the same way. □

8.8. Corollary. Let R be a filtered ring such that FR is discrete and exhaustive. Let $N \in R$-filt, $M \in$ filt-R be such that FM and FN are discrete and exhaustive, then $\mathrm{Tor}_n^{G(R)}(G(M),G(N)) = o$ implies : $\mathrm{Tor}_n^R(M,N) = o$.

Proof. Consider a free resolution of N :

$$\cdots \longrightarrow L_2 \xrightarrow{f_2} L_1 \xrightarrow{f_1} L_0 \xrightarrow{f_0} N \longrightarrow o \ ,$$

where F_iL is discrete and exhaustive, for all i, and where $f_0, f_1, f_2, \ldots$ are strict morphism. Consider the following commutative diagram, the vertical arrows of which represent isomorphisms :

$$\begin{array}{ccccccccc}
\cdots \longrightarrow & G(M) \underset{G(R)}{\otimes} G(L_2) & \longrightarrow & G(M) \underset{G(R)}{\otimes} G(L_1) & \longrightarrow & G(M) \underset{G(R)}{\otimes} G(L_0) & \longrightarrow & o \\
 & \downarrow & & \downarrow & & \downarrow & & \\
\cdots \longrightarrow & G(M \underset{R}{\otimes} L_2) & \longrightarrow & G(M \underset{R}{\otimes} L_1) & \longrightarrow & G(M \underset{R}{\otimes} L_0) & \longrightarrow & o
\end{array}$$

We have : $G_\ell^{\star,n}(M \underset{R}{\otimes} L_\star) = H_n G(M \underset{R}{\otimes} L_\star) = H_n(G(M) \underset{G(R)}{\otimes} G(L_\star)) = \mathrm{Tor}_n^{G(R)}(G(M),G(N))$.
If $\mathrm{Tor}_n^{G(R)}(G(M),G(N)) = o$ then $G^{\,,n}(M \underset{R}{\otimes} L_\star) = o$ and, since $F(M \underset{R}{\otimes} L_\star)$ is exhaustive and discrete, $\mathrm{Tor}_n^R(M,N) = H_n(M \underset{R}{\otimes} L_\star) = o$. □

Special References for Chapter I.

K.S. Brown, F. Dror [4]

R. Fossum, H.B. Foxby [8]

P. Gabriel, Y. Nouaze [27]

C. Nastasescu [28]

G. Sjöding [35]

CHAPTER II. TOPICS IN GRADED RING THEORY.

II.1. HOMOGENIZATION.

Let $R = \bigoplus_{i\in\mathbb{Z}} R_i$ be a graded ring. If $M\in R$-gr then we put $M^+ = \bigoplus_{i\geq 0} M_i$, $M^- = \bigoplus_{i\leq 0} M_i$. It is obvious that R^+ and R^- are graded subrings of R, whereas M^+ is a graded R^+-module, M^- is a graded R^--module. If $N\subset M$ are graded R-modules then we have $(M/N)^+ = M^+/N^+$ and $(M/N)^- = M^-/N^-$.

Let $M\in R$-gr and let $X\subset \underline{M}$ be an R-submodule. Any $x\in X$ may be, in a unique way, written as $x_1 + \dots + x_n$ with $\deg x_1 < \dots < \deg x_n$. By $X^\sim$ (resp. $X_\sim$) we will denote the submodule of $\underline{M}$ generated by the x_n (resp. x_1) i.e. by the homogeneous components of highest (resp. lowest) degree of elements of X. With these notations we have :

1.1. Lemma. 1. $X^\sim$ and $X_\sim$ are graded submodules of M.

2. $X = X^\sim$ if and only if X is a graded submodule of M.
3. If $X\subset Y$ are R-submodules of $\underline{M}$ then $X^\sim\subset Y^\sim$, $X_\sim\subset Y_\sim$.
4. Equivalently : 1°. $X = o$, 2°. $X^\sim = o$, 3°. $X_\sim = o$.
5. If M is left-or right-limited and if $X^\sim$ (resp. $X_\sim$) is generated by r homogeneous elements then X may be generated by less than r elements.
6. If I is a left-ideal of R and N an R-submodule of M, then : $I^\sim N^\sim \subset (IN)^\sim$ and $I_\sim N_\sim \subset (IN)_\sim$.
7. If I is an ideal of R then by rad I we mean $\cap\{P$, P a prime ideal of R, $P\supset I\}$. If either R is commutative or left Noetherian, we have : $(\text{rad } I)^\sim \subset \text{rad}(I^\sim)$ and $(\text{rad } I)_\sim \subset \text{rad}(I_\sim)$.

Proof. Assertions 1,2,3,4 are easy to prove.

5. Filter X as follows $F_n X = X \cap \bigoplus_{i\geq -n} M_i$. It is easily verified that $X^\sim$ is isomorphic to the graded module G(X) associated to the filtered module X. Similarly $X_\sim$ is isomorphic to G(X) when X is filtered by putting $F_n X = X\cap \bigoplus_{i\leq n} M_i$. The statement now follows directly from Proposition I.5.3.

6. Let $\tilde{a}\in I^\sim$, $\tilde{x}\in N^\sim$ be homogeneous elements. Then there exist $a\in I, x\in N$ such that $\tilde{a},\tilde{x}$ are the homogeneous components of highest degree of a,x resp. If $\tilde{a}\tilde{x}\neq o$ then $\tilde{a}\tilde{x}$

is the homogeneous component of highest degree of $ax \in IN$, hence $\widetilde{\widetilde{ax}} \in (IN)^{\sim}$.

7. The commutative case is easy and, as a matter of fact, it may be proven exactly in the same way as in the following proof for left Noetherian R. Take $\tilde{x} \in (\text{rad } I)^{\sim}$. First note that for any ideal J of R, $J^{\sim}$ and $J_{\sim}$ are ideals of R. Suppose that $\bar{x}$ is the component of heighest degree of $x \in \text{rad } I$. Since R is left Noetherian $x \in \text{rad } I$ is equivalent to $R_{\lambda_1} x R_{\lambda_2} x \dots R_{\lambda_{n-1}} x R_{\lambda_n} \subset I$, for all $(\lambda_1, \dots, \lambda_n) \in \mathbb{Z}^n$, for some $n \in \mathbb{Z}$. Taking components of highest degree yields : $R_{\lambda_1} \tilde{x} R_{\lambda_2} \tilde{x} \dots R_{\lambda_{n-1}} \tilde{x} R_{\lambda_n} \subset \tilde{I}$ (note that there are no problems arising from zero products), for all $(\lambda_1, \dots, \lambda_n) \in \mathbb{Z}^n$.

1.2. Proposition. Let $X \subset Y$ be submodules of $\underline{M}$, then the following conditions are equivalent :

1. $X = Y$.
2. $X^{\sim} = Y^{\sim}$ and $X \cap M^{-} = Y \cap M^{-}$.
3. $X_{\sim} = Y_{\sim}$ and $X \cap M^{+} = Y \cap M^{+}$.

Proof. In order to prove that $1 \Leftrightarrow 2$ let $y \in Y$, $y = y_1 + \dots + y_m$ the homogeneous decomposition of y, and let $\deg y_m = t$. We shall show that $y \in X$ by induction on t. If $t \leqslant o$ then $y \in M^{-}$ and thus $y \in X$. If $t > o$ then $y_m \in Y^{\sim} = X^{\sim}$ yields that there is an $x \in X$ such that $x = x_1 + \dots + x_{n-1} + y_m$, with $\deg x_1 < \dots < \deg x_{n-1} < t$. It follows that $y - x \in Y$ has a homogeneous decomposition in which the highest degree appearing is less than t. The induction hypothesis yields that $y - x \in X$ and hence $y \in X$. The implications $1 \Leftrightarrow 3$ are established in a similar way. □

1.3. Corollary. With notations as above : if M is left-limited (resp. right-limited) then $X = Y$ if and only if $X^{\sim} = Y^{\sim}$ (resp. $X_{\sim} = Y_{\sim}$).

If R is a graded ring and $\underline{M} \in R\text{-mod}$ then it is not always possible to consider maximal graded submodules within M; for special $\underline{M}$ however we have the following:

1.4. Lemma. Let $N \in R\text{-gr}$ and let $\underline{M}$ be a submodule of $\underline{N}$. There is a unique object in R-gr which is maximal amongst objects of R-gr which are submodules of M and graded submodules of N, this object will be denoted by $(M)_g$.

Proof. cf. [30].

1.5. Lemma. Let R be a graded ring. If I is an ideal of R then $(I)_g$ is a graded ideal of R. If P is any prime ideal of R then $(P)_g$ is a graded prime ideal of R. If J is a graded ideal of R then $\text{rad}J = \cap\{(P)_g, P \text{ a prime ideal containing } J\}$, hence rad J is a graded ideal of R too.

1.6. Note. $(M)_g$ as defined in 1.4 depends on the structure of N, however we will only use this lemma in cases where all modules considered are submodules of some fixed graded module, so there is no need to take this dependence into account in the notation; e.g. in Lemma 1.5. all gradations considered are induced by the gradation of the ring R.

II.2. THE STRUCTURE OF PRINCIPAL GRADED RINGS.

In this section we reduce the study of principal graded rings to the study of twisted polynomial rings, a class of rings which will also reappear in II Throughout this section $R = R_o \oplus R_1 \oplus \dots$ will be a positively graded domain.

2.1. Lemma. Let $I \subset R$ be a left principal graded ideal, then there exists a homogeneous element which generates it.

Proof. Suppose $I = Rx$ and write $x = x_1 + \dots + x_n$ with $\deg x_1 < \dots < \deg x_n$. Since $x_j \in I$ for all j we have that $x_2 = rx$ with $r \in R$, $r \neq o$. Write $r = r_1 + \dots + r_m$ with $\deg r_1 < \dots < \deg r_m$. Since $r_1x_1 \neq o$, $x_2 = r_1x_1$ follows. Repeating this argumentation we find that $x_i = r_ix_1$ for some homogeneous $r_i \in R$, therefore $x = (1 + r_1 + \dots + r_{n-1})x_1$ and $Rx = Rx_1$ follows. □

2.2. Theorem. In case the gradation is nontrivial the following statements are equivalent :

1. R is a left principal ring.

2.a. R_o is a left principal ring.

b. There is an injective morphism $\varphi : R_o \to R_o$ mapping nonzero elements of R_o to units of R_o

c. $R \cong R_o[X,\varphi]$ as graded rings, where $\deg x > o$.

Proof. $1 \Rightarrow 2$. If $I_0 \subset R_0$ is a left ideal then RI_0 is graded and left principal, therefore, by the foregoing lemma there is a homogeneous $a \in R$ such that $RI_0 = Ra$. If $I_0 \neq 0$ then the fact that the gradation is positive yields that $a \in R_0$ and $(RI_0)_0 = I_0 = R_0 a$; thus R_0 is left principal. Now to prove b), consider the ideal $R_+ = \bigoplus_{i \geq 1} R_i$ which is nonzero as the gradation is nontrivial. By our hypothesis $R_+ = Rt$ for some homogeneous $t \in R$ say $\deg t = s \geq 1$ i.e. $A \in R_s$. From $R_+ = Rt$ it follows that $R_s = R_0 t$, $R_{2s} = R_0 t^2$ ..., while $R_i = 0$ if s i. Pick $a \in R_0$; then $ta \in Rt$ since Rt is two-sided, hence there exists a unique element $\varphi(a) \in R$ such that $ta = \varphi(a)t$, clearly $\varphi(a) \in R_0$. It is easy to verify that φ, thus defined, is an injective ring homomorphism $R_0 \to R_0$. Again consider $a \in R_0, a \neq 0$. The left ideal $R_0 a + R_+$ is principal, hence there exists a $b \in R_0$, $b \neq 0$, such that $R_0 a + R_+ = Rb$. In this case : $R_0 a = R_0 b, \ldots, R_{ns} = R_{ns} b$. Substituting $R_s = R_0 t$ we get : $R_0 t = R_0 tb = R_0 \varphi(b)t$ and so there exists $\lambda \in R_0$ such that $t = \lambda\varphi(b)t$, hence $\lambda\varphi(b) = 1$. From $R_0 a = R_0 b$ we derive : $R_0 \varphi(a) = R_0 \varphi(b) = R_0$, thus $\mu\varphi(a) = 1$ for some $\mu \in R_0$. Applying φ yields $\varphi(\mu)\varphi^2(a) = 1$. But $\varphi(\mu)$ has a left-inverse μ', obtained as before, so $\mu'\varphi(\mu) = 1$ yields $\mu' = \varphi^2(a)$ or $\varphi^2(a)\varphi(\mu) = 1$. Since φ is injective it results from this that $\varphi(a)\mu = 1$, consequently $\varphi(a)$ is a unit of R_0.

For c) it suffices to check that the map $R_0[X,\varphi] \to R$ defined by : $a \to a$ if $a \in R_0$, $X \to t$, actually defines an isomorphism of graded rings.

$2 \Rightarrow 1$. It is clearly sufficient to prove that in $R_0[X,\varphi]$; any graded left ideal is left principal and without loss of generality we may assume that $\deg X = 1$. Every left graded ideal I of $R_0[X,\varphi]$ has the form $I = I_0 \oplus I_1 X \oplus I_2 \text{-} X^2 \oplus \ldots$, where $I_0, I_1, I_2, \ldots$, are left ideals of R_0 such that $I_n \subset \varphi^{-1}(I_{n+1})$ for each $n \geq 0$. Let $r \in \mathbb{N}$ be minimal such that $I_r \neq 0$, then, since $I_r \subset \varphi^{-1}(I_{r+1})$, I_{r+1} contains units of R_0 or $I_{r+1} = R_0$. We reached the situation $I_n = R_0$ for all $n > r$. If $I_r = R_0 b$ then $I = Rb$ follows, consequently R is a left principal ring. □

2.3. Corollary. The following conditions are equivalent :

1. R is left and right principal.

2.a. R_0 is a skewfield.

b. There is an automorphism $\varphi : R_0 \to R_0$ such that $R \cong R_0[X,\varphi]$ as graded rings, where deg $x > o$.

Proof. $2 \Rightarrow 1$. Trivial.

$1 \Rightarrow 2$. Since R is left principal, $R \cong R_0[X,\varphi]$, where φ is a monomorphism $R_0 \to R_0$. The right ideal $R_0 \oplus \varphi(R_0)X \oplus \varphi^2(R_0)X^2 \oplus \dots$ has the form aR for some $a \in R_0$. It follows that $R_0 = aR_0$, $\varphi(R_0)X = aR_1 = aR_0X$ and so we obtain that $\varphi(R_0) = aR_0 = R_0$ i.e. φ is an automorphism. Since φ maps nonzero elements of R_0 onto units of R_0, it follows that R_0 is a skewfield.

2.4. Remark. In the situation of 2.3. and by construction of φ it is clear that φ^n is inner if and only if there exists $b \neq o$ in R_0 such that bt^n commutes with R_0. Let K_0 be the center of R_0 and let $k \subset K_0$ be the fixed field for φ. If the center C of R properly contains k then $C \cong k[bt^n]$ for some $n \in \mathbb{N}$, $b \in R_0 - \{o\}$, and in this case φ^n is an inner automorphism. Conversely if φ^n is inner (n the smallest integer as such) then $C \cong k[bt^n]$ for some $b \in R_0 - \{o\}$. If φ^n is not inner for any $n \in \mathbb{N}$ then $C = k$ and all ideals of R are generated by powers of t. If φ^n is inner for some $n \in \mathbb{N}$ then all ideals of R not in R_+ may be generated by a central element. Note also that R is a noncommutative graded Dedekind domain.

2.5. Remark. If K is a field then the only possible $\mathbb{Z}$-gradation on K is the trivial gradation. Indeed if $x \neq o \in K_i$ with $i > o$ then expressing $y = (1+x)^{-1} = \sum_{j \in \mathbb{Z}} y_j$ yields conditions : $1 = y_0 + xy_{-i}$, $o = y_j + xy_{j-1}$ for all $j \neq o$. So if n_0 is minimal such that $y_{n_0} \neq o$ and if $n_0 \neq o$ then the second relation yields $y_{n_0} + xy_{n_0 - i} = o$ i.e. $y_{n_0} = o$, contradiction. If $n_0 = o$ then one calculates $y = 1 - x + x^2 - \dots$ but as $x^n \neq o$ for all n this is impossible. However we shall introduce the notion of graded division ring, (some say:graded field), this is a graded ring such that each nonzero homogeneous element is invertible. For example, let k be a field and consider $K = k[X,X^{-1}]$ for some variable X, then K, equipped with the obvious grading, is a graded division ring. Now consider $K[Y] = R$ and grade it as follows $R_n = \sum_{i+j=n} K_i Y^j$. An $f \in R_n$ is written : $f = a_0 + a_1Y + \dots + a_kY^k$ where $a_0 \in K_n, \dots, a_k \in K_{n-k}$. If $g \in R_m$ and $g \neq o$, then there exist homogeneous elements q,r in R such that $f = gq + r$ with $\deg_Y r < \deg_Y g$. Having a division algorithm on R it follows that graded ideals of R are generated by one homogeneous element of R. However R is not a (left) principal ring as is well-known (since K is not a field!), this shows that the condition "R is positively graded" cannot be dropped from II.2.

II.3. NOETHERIAN OBJECTS.

Let $M \in R$-gr. Then M is said to be a left Noetherian (Artinian) object, and in this case we say that M is gr.ℓ.Noetherian (gr.ℓ.Artinian) if M satisfies the ascending chain condition (descending chain condition) for graded submodules of M. It is straightforward to verify that M is gr.ℓ.Noetherian if and only if each graded submodule of M is finitely generated , or if and only if each non-empty family of graded submodules of M has a maximal element. Dually, M is gr.ℓ.Artinian if and only if each non-empty family of graded submodules of M has a minimal element, or if and only if each intersection of graded submodules may be reduced to a finite intersection.

3.1. Proposition. Let $M \in R$-gr have left-(right-) limited grading. Then M is gr.ℓ.Noetherian, resp.gr.ℓ.Artinian, if and only if $\underline{M}$ is ℓ.Noetherian, resp. ℓ.Artinian, in R-mod.

Proof. See Corollary 1.3.

Without the left-limitedness condition the statement remains valid in the Noetherian case as we will see later, but it fails for the Artinian case. Indeed, let K be a graded division ring, then K is clearly gr.ℓ.Artinian but clearly $\underline{K}$ need not be ℓ.Artinian as the example $K = k[X,X^{-1}]$ shows.

3.2. Proposition. Let R be a graded ring, $M \in R$-gr a left Noetherian object, then :

1. For all $i \in \mathbb{Z}$, M_i is a left Noetherian R_o-module.
2. M^+ is a left Noetherian R^+-module.
3. M^- is a left Noetherian R^--module.

Conversely if M^+,M^- are left Noetherian objects of R^+-mod, R^--mod resp. then M is gr.ℓ.Noetherian.

Proof. 1. Let N_i be any R_0-submodule in M_i. Then RN_i is a graded submodule in M and hence it is finitely generated by homogeneous elements $x_1,\dots,x_k$, which may be supposed to be taken from N_i. Pick $y\in N_i$; then there exist homogeneous $\lambda_1,\dots,\lambda_k\in R$ such that $y=\sum_{i=1}^{k}\lambda_i x_i$. Comparision of degrees yields $\deg\lambda_j=0$, $j=1,\dots,k$ and thus N_i is generated as an R_0-module by the elements $x_1,\dots,x_k$.

2. By Proposition 3.1. it is sufficient to show that M^+ is gr.ℓ.Noetherian in R^+-gr. Let $N=\bigoplus_{i\geq 0} N_i$ be a graded submodule of M^+. As in 1. we may assume that RN is generated by homogeneous elements $x_1,\dots,x_k$ of N. Put $t=\max\{\deg x_1,\dots,\deg x_k\}$ and let $y\in N$ be a homogeneous element with $\deg y\geq t$. There exist $\lambda_1,\dots,\lambda_k\in h(R)$ such that $y=\lambda_1x_1+\dots\lambda_kx_k$. Now, $\deg y\geq t$ implies that $\deg\lambda_i\geq 0$ for all $i=1\dots k$, consequently $\lambda_i\in R^+, i=1\dots k$. Because of 1., $M_0\oplus\dots\oplus M_{t-1}$ is a left Noetherian R_0-module and therefore $N_0\oplus\dots\oplus N_{t-1}$ is a finitely generated R_0-module. Let $y_1,\dots,y_s$ generate $N_0\oplus\dots\oplus N_{t-1}$ over R_0. Then clearly $\{x_1,\dots,x_k,y_1,\dots,y_s\}$ generates N over R^+, therefore M^+ is a left Noetherian R^+-module.

3. Similar to 2.

Supposing that M^+ and M^- are both left Noetherian objects in R^+-mod, R^--mod resp., let $N_1\subset N_2\subset\dots\subset N_p\subset\dots$ be an ascending chain of graded submodules of M. Then we also obtain ascending chains in M^+ and M^- :

$$N_1^+\subset N_2^+\subset\dots\subset N_p^+\subset\dots\subset M^+$$

$$N_1^-\subset N_2^-\subset\dots\subset N_p^-\subset\dots\subset M^- .$$

Our assumptions amount to : $N_k^+=N_{k+1}^+$ and $N_k^-=N_{k+1}^-$, but then $N_k=N_{k+1}=\dots$. □

3.3. Theorem. Let $M\in R$-gr. then the following assertions are equivalent :

1. M is gr.ℓ.Noetherian.
2. $\underline{M}$ is a ℓ.Noetherian $\underline{R}$-module.

Proof. Since 2⇒1 is obvious let us check 1⇒2. Let $X_1\subset X_2\subset\dots\subset X_n\subset\dots$ be an ascending chain of submodules of $\underline{M}$. By 3.2 there is an $n_0\in\mathbb{N}$ such that $M^-\cap X_i=M^-\cap X_{i+1}$ and $\tilde{X}_i=\tilde{X}_{i+1}$ for each $i\geq n_0$. By Proposition 1.2. : $X_i=X_{i+1}=\dots$, $i\geq n_0$. □

3.4. Proposition. Let R be a graded ring, $M \in R\text{-gr}$. Suppose there exist elements $a \in R_1$, $b \in R_{-1}$ such that :

1. $R_0 a = aR_0$ and $R_0 b = bR_0$.
2. $M_i = a^i M_0$ and $M_{-i} = b^i M_0$ for all $i \geq 0$.

If M_0 is a left Noetherian R_0-module then M is a left Noetherian R-module.

Proof. By Proposition 3.2. and 3.1. we may reduce the proof to the case of positively graded R and M. Let $N = \bigoplus_{i \geq 0} N_i$ be a graded submodule of M. For each $i \geq 0$ put : $P_i = \{x \in M_0,\ a^i x \in N_i\}$. Obviously, each P_i is an R_0-submodule of M_0 and $P_0 = N_0$, $P_i \subset P_{i+1}$ for all i. Also, $a^i P_i = N_i$ for all $i \geq 0$. Now, M_0 being a left-Noetherian R_0-module, there is an $n \in \mathbb{N}$ such that : $P_n = P_{n+1} = \dots$. Choose $\{z_0^{(1)},\dots,z_0^{(r_0)}\}$ to be a system of generators for $P_0 = N_0$ and let $\{z_i^{(1)},\dots,z_i^{(r_i)}\}$ generate P_i, $i = 1,\dots,n$. Note that the set :

$$\{z_0^{(1)},\dots,z_0^{(r_0)}, az_1^{(1)},\dots,az_1^{(r_1)},\dots,a^n z_n^{(1)},\dots,a^n z_n^{(r_n)}\}$$

generates N as a left R-module. □

3.5. Corollary Let R be a graded ring and suppose there exist elements $a \in R_1$, $b \in R_{-1}$ such that :

1. $R_1 = R_0 a = aR_0$ and $R_{-1} = R_0 b = bR_0$.
2. $R_1 R_i = R_{i+1}$ and $R_{-1} R_{-i} = R_{-i-1}$ for all $i \geq 0$.

If M is a left-Noetherian R_0-module then $R \otimes_{R_0} M$ is a left Noetherian R-module.

Proof.

$$R \otimes_{R_0} M = \bigoplus_{i \in \mathbb{Z}} R_i \otimes_{R_0} M \cong \bigoplus_{i > 0} (a^i \otimes_{R_0} M) \oplus M \oplus (\bigoplus_{i < 0} b^i \otimes_{R_0} M) \cong \bigoplus_{i > 0} a^i M \oplus M \bigoplus_{i < 0} b^i M$$

Now apply Proposition 3.4. □

3.6. Corollary. Let R be a graded ring as in 3.5. If R_0 is a left Noetherian ring then R is left Noetherian.

3.7. Corollary. Let R be a left Noetherian ring. Let $\varphi : R \to R$ be a ring automorphism and let $\delta : R \to R$ be a φ-derivation of R. Then $S = R[X,\varphi,\delta]$ is left Noetherian.

Proof. Filter S by $F_i S = \{P \in S, \deg P \leq i\}$; this filtration is exhaustive and discrete. The associated graded ring G(R) is isomorphic to $R[X,\varphi]$. Because of Corollary 5.6. it will be sufficient to show that $R[X,\varphi]$ is left Noetherian, but this follows directly from Corollary 3.6. □

Let R and S be any two rings and let $M \in$ R-mod-S, $N \in$ S-mod-R be bimodules. Consider the matrix ring $T = \begin{pmatrix} R & M \\ N & S \end{pmatrix} = \left\{ \begin{pmatrix} r & m \\ n & s \end{pmatrix}, r \in R, m \in M, n \in N, s \in S \right\}$ with usual addition, and multiplication defined by : $\begin{pmatrix} r & m \\ n & s \end{pmatrix} \begin{pmatrix} r' & m' \\ n' & s' \end{pmatrix} = \begin{pmatrix} rr' & rm'+ms' \\ nr'+sn' & ss' \end{pmatrix}$.
This ring has gradation $T_{-1} = \begin{pmatrix} 0 & 0 \\ N & 0 \end{pmatrix}$, $T_0 = \begin{pmatrix} R & 0 \\ 0 & S \end{pmatrix}$ and $T_1 = \begin{pmatrix} 0 & M \\ 0 & 0 \end{pmatrix}$, $T_i = 0$ if $|i| > 1$. The left graded ideals of T are of the form $\begin{pmatrix} I & M' \\ N' & J \end{pmatrix}$ where I and J are left ideals in R and S resp., while M' is an R-submodule of M, N' an S-submodule of N, such that $MJ \subset M'$, $NI \subset N'$.

3.8. Corollary. T is left Noetherian (Artinian) if and only if R and S are left Noetherian (Artinian), M is a finitely generated R-module, N a finitely generated S-module.

Let $R = \bigoplus_{i \in \mathbb{Z}} R_i$ be a graded ring, $M = \bigoplus_{i \in \mathbb{Z}} M_i$ a graded R-module. For any $(d,k) \in \mathbb{Z}^2$ such that $d \geq 1$, $0 \leq k \leq d-1$ we put :

$$R^{(d)} = \bigoplus_{i \in \mathbb{Z}} R_{id} \quad \text{and} \quad M^{(d,k)} = \bigoplus_{i \in \mathbb{Z}} M_{id+k} \ .$$

Clearly, $R^{(d)}$ is a graded subring of R and, $M^{(d,k)}$ is a graded $R^{(d)}$-module. If N is a graded submodule of M then $N^{(d,k)}$ is a graded submodule of $M^{(d,k)}$. We shall write $M^{(d)}$ instead of $M^{(d,o)}$ for every $d \geq 1$. For fixed d, it is clear that M is the direct sum of the $R^{(d)}$-modules $M^{(d,k)}$ with $0 \leq k \leq d-1$.

3.9. Proposition. With notations as above, the following statements are equivalent:
1. M is a left Noetherian R-module.
2. $M^{(d,k)}$ is a left Noetherian graded $R^{(d)}$-module for every $1 \leq k \leq d-1$.

Proof. $1\Rightarrow 2$. Let $P\subset M^{(d,k)}$ be a graded submodule, then RP is a graded submodule of M and hence it is finitely generated over R, let $x_1,\dots,x_n$ be homogeneous elements of R which generate RP. If $y\in h(P)$ then there exist homogeneous $\lambda_1,\dots,\lambda_n\in R$ such that $y=\lambda_1x_1+\dots+\lambda_nx_n$. Put : $\deg x_1=i_1d+k,\dots,\deg x_n=i_nd+k$, $\deg y=id+k$. It follows that : $\deg\lambda_1=(i-i_1)d,\dots,\deg\lambda_n=(i-i_n)d$, hence $\lambda_1,\dots,\lambda_n\in R^{(d)}$. So P is generated by $x_1,\dots,x_4$ over $R^{(d)}$.

$2\Rightarrow 1$. Let N be a graded submodule of M. Since $N^{(d,k)}$ is then a graded $R^{(d)}$-submodule of $M^{(d,k)}$ it may be generated by : $x_1^{(k)},\dots,x_{n_k}^{(k)}$, over $R^{(d)}$. However, as $N=\bigoplus_{o\leqslant k\leqslant d-1}N^{(d,k)}$, it follows that $\bigcup_{o\leqslant k\leqslant d-1}\{x_1^{(k)},\dots,x_{n_k}^{(k)}\}$ generates R over R. □

3.10. Remark. Let P be a graded $R^{(d)}$-submodule of $M^{(d,k)}$, then $RP\cap M^{(d,k)}=P$.

3.11. Corollary. If R is a left Noetherian graded ring then the ring $R^{(d)}$ $(d\geqslant 1)$ is left Noetherian.

Another type of graded left Noetherian rings is given by the construction of the Rees ring associated to certain ideals. Let R be a ring, I an ideal of R. The Rees ring associated to I is the ring :

$$R(I)=R+IX+I^2X^2+\dots+I^nX^n+\dots\subset R[X],$$

which is obviously a graded subring of $R[X]$ isomorphic to $R\oplus I\oplus I^2\oplus\dots\oplus I^n\oplus\dots$. Moreover $R(I)/I\oplus I^2\oplus\dots\oplus I^n\oplus\dots\cong G(I)=R/I\oplus I/I^2\oplus\dots\oplus I^n/I^{n+1}\oplus\dots$.

3.12. Proposition. Let R be a left Noetherian ring and let the ideal I of R be generated on the left by a central system (cf. I.5.12) : $\{a_1,\dots,a_n\}$, such that the a_i, $i=1..n$, commute amongst themselves, then R(I) is a left Noetherian ring.

3.13. Sublemma. Let S be a left Noetherian subring of an arbitrary ring R. Suppose that R is generated as a ring by S and a single element x such that $sx-xs\in S$ for all $s\in S$, then R is a left Noetherian ring.

Proof of the sublemma : Clearly $R = S[x] = \{a_n x^n + \dots + a_1 x + a_0, a_i \in S, i = 0\dots n\}$. We equip R with a discrete filtration FR as follows : for $t \in \mathbb{N}$, $F_t R = \{a_n x^n + \dots + a_0, a_i \in S\ i = 0,\dots,n, n \leq t\}$, in particular $F_0 R = S$. For any $n \in \mathbb{N}$, $a \in S$ we have $ax^n - x^n a = b_{n-1} x^{n-1} + \dots + b_0$ for some $b_j \in S$, $j = 0,\dots,n-1$. Hence $ax^n - x^n a \in F_{n-1} R$. Now let G(R) be the graded ring associated to the filtered ring R, $G(R) = F_0 R \oplus F_1 R / F_0 R \oplus \dots \dots \oplus F_n R / F_{n-1} R \oplus \dots$. In this ring, the image $\overline{x}$ of x in $F_1 R / F_0 R$ is a central element, thus there exists a ring epimorphism $\psi : S[Y] \to G(R)$, $\psi(Y) = \overline{x}$. Since S[Y] is left Noetherian, G(R) is left Noetherian too, but then R is also left Noetherian because of Corollary I.5.7. □

Proof of the proposition. Put $I_k = (a_1,\dots,a_k)$ for $k \leq n$. Since $a_1 \in Z(R)$, $R(I_1)$ is left Noetherian. We proceed by induction. Since $R(I_n)$ is generated as a ring by $R(I_{n-1})$ and the single element $x = a_n X$ (this is readily checked) we try to apply the sublemma. This will be possible because for all $z \in R(I_{n-1})$ we have that $a_n Xz - za_n X \in R(I_{n-1})$; indeed, this follows immediately from the fact that the image of a_n in $R/(a_1,\dots,a_{n-1})$ is in the center of the latter ring (also making further use of the fact that a_n commutes with $a_1,\dots,a_{n-1}$). □

3.14. Remark. The conditions of the proposition are fulfilled if I is generated by a central system (a_1,a_2) or in case I is generated by central elements of R.

Let I be an ideal of a ring R, and consider R as a filtered ring with the I-adic filtration. Consider an $M \in R$-filt. We say that the filtration FM is I-fitting if there exists a $p \in \mathbb{Z}$ such that $IF_n M = F_{n-1}$ for all $n \leq p$. As an example $F_{-i} M = I^i M$, $i \in \mathbb{Z}$, defines an I-fitting filtration. (Note that $F_n M = M$ for all $n \geq 0$). If FM is an I-fitting filtration then the quotient filtration on M/N, for any submodule N of M, is also I-fitting. If FM is I-fitting then $R(M) = M_0 \oplus M_{-1} \oplus M_{-2} \oplus \dots \dots \oplus M_{-n} \oplus \dots$ is the Rees module associated to M. It is clear that R(M) is a graded R(I)-module. The I-fitting filtrations, for certain I, are characterized by the following,

3.15. Proposition. Let R be a left Noetherian ring and let I be an ideal of R generated by a central system. Let R have the I-adic filtration and let $M \in R$-filt be finitely generated. Then, the following statements are equivalent : 1°. FM is I-fitting, 2°. R(M) is an R(I)-module of finite type.

Proof.: If FM is I-fitting then there exists a $p \in \mathbb{Z}$ such that for all $n \leq p$: $IM_n = M_{n-1}$. Consequently R(M) will be generated as a left R(I)-module by the set $M_0 \oplus M_{-1} \oplus \dots \oplus M_p$, which is however finitely generated as an R-module. Conversely, suppose that R(M) may be generated by elements $x_1, \dots, x_n$ of degree $d_i = \deg x_i$, $i = 1, \dots, n$. Putting $p = \max\{d_i, i = 1, \dots, n\}$ one easily establishes that $M_{p-1} = I M_p$. □

3.16. Proposition. Let R be a left Noetherian ring and let I be an ideal of R generated by a commuting central system. Let R have the I-adic filtration and $M \in R$-filt be finitely generated. If FM is I-fitting then, for every submodule N of M, the filtration FN given by $F_i N = F_i M \cap N$ is I-fitting.

Proof. A straightforward combination of Proposition 3.12 and Proposition 3.15. □

3.17. Remark. Proposition 3.16 may be considered as a generalization of the Artin-Rees property (however only ideals generated by commuting central systems have been considered here!).

II.4. KRULL DIMENSION OF GRADED RINGS

In Section I.5 we have introduced the notion of Krull dimension in R-Mod as well as in R-gr. Let us first state some easy lemmas and then turn to relating $K.\dim_R(M)$ and $K.\dim_{R\text{-gr}}(M)$ for any $M \in R$-gr.

4.1. Lemma. Let R be a graded ring, $M \in R$-gr. If N is a graded submodule of M then M has Krull dimension if and only if N and M/N have Krull dimension. In this case

$$K.\dim_{R\text{-gr}} M = \sup(K.\dim_{R\text{-gr}} N, K.\dim_{R\text{-gr}} M/N).$$

Proof. Similar to the ungraded case, cf. Lemma 1.1. in [12]. □

4.2. Lemma. Let $M \in R\text{-gr}$ be a left Noetherian object then M has Krull dimension.

Proof. Similar to Proposition 1.3. in [12]. □

4.3. Lemma. Let R be a graded ring, $M \in R\text{-gr}$. Suppose that M has Krull dimension and let α be equal to $\sup\{1+\text{K.dim}_{R\text{-gr}} M/N$, N an essential graded submodule of $N\}$. Then we have : $\alpha \geq \text{K.dim}_{R\text{-gr}} M$.

Proof. Similar to Corollary 1.5. in [12].

4.4. Lemma. Let R be a graded ring and let $M \in R\text{-gr}$ be left- or right- limited, then:

1. M has Krull dimension if and only if $\underline{M}$ has Krull dimension, in case this happens we have that $\text{K.dim}_{R\text{-gr}} M = \text{K.dim}_R M$.
2. If M is α-critical then $\underline{M}$ is α-critical.

Proof. 1. Follows from Corollary 1.3.

2. Suppose $M = \bigoplus_{i \in \mathbb{Z}} M_i$ where $M_i = o$ for all $i < n_o$. Let $X \neq o$ be a submodule of $\underline{M}$, then $\text{K.dim}_{R\text{-gr}} M/X^{\sim} < \alpha$. By Proposition 1.2., Corollary 1.3. and also Lemma I.5.4., we have $\text{K.dim}_{\underline{R}} \underline{M}/X \leq \text{K.dim}_{R\text{-gr}} M/X^{\sim}$. Therefore $\underline{M}$ is also α-critical.

4.5. Proposition. Let R be a positively graded ring and let $M \in R\text{-gr}$. Then the following properties hold :

1. M has Krull dimension if and only if $\underline{M}$ has Krull dimension, and if so, we have $\text{K.dim}_{R\text{-gr}} M = \text{K.dim}_{\underline{R}} \underline{M}$.
2. If M is α-critical then $\underline{M}$ is α-critical.

Proof. 1. Let $M = \bigoplus_{i \in \mathbb{Z}} M_i$. For each $p \in \mathbb{Z}$ put $M_{>p} = \bigoplus_{n>p} M_n$. Obviously $M_{>p}$ is a graded submodule of M and also $(M/M_{>p})_i = o$ for every $i > p$. The foregoing lemma yields $\text{K.dim}_{R\text{-gr}} M_{>p} = \text{K.dim}_R \underline{M_{>p}}$ and $\text{K.dim}_{R\text{-gr}} M/M_{>p} = \text{K.dim}_{\underline{R}} \underline{M/M_{>p}}$. Application of Lemma 4.1 enables us to deduce 1.

2. Let $X \neq o$ be a submodule of $\underline{M}$. For some $p \in \mathbb{Z}$, $X \cap M_{>p} \neq o$. Since M is α-critical and the foregoing lemma then implies that $\underset{R\text{-gr}}{\text{K.dim}}\ M_{>p}/(X \cap M_{>p})$ is smaller than α. Clearly, there exists a strictly increasing mapping from the lattice of submodules of $\underline{M}/X$ into the set-product of the lattices of submodules of $M_{>p}/X \cap M_{>p}$ and $M/M_{>p}$ respectively; namely :

$$Y/X \longrightarrow (Y \cap M_{>p})/(X \cap M_{>p}), (Y + M_{>p})/M_{>p} .$$

Now $\underset{R}{\text{K.dim}}\ \underline{M}/\underline{M}_{>p} < \alpha$, hence from Lemma I.5.4. we derive that $\underset{R}{\text{K.dim}}\ \underline{M}/\underline{X} < \alpha$ or $\underline{M}$ is α-critical. □

4.6. Lemma. Let R be a left Noetherian graded ring and let $M \in R\text{-gr}$ be a uniform object. Suppose either that M is limited or that the gradation of R is positive. Then $\underline{M}$ is a uniform $\underline{R}$-module.

Proof. Since R is left Noetherian, M contains a nonzero graded submodule N which is α-critical (cf. Theorem 2.7., [12]). Uniformity of M yields that M is an essential extension of N in R-gr; but by Lemma I.3.3.13. it follows then that $\underline{M}$ is an essential extension of $\underline{N}$. Lemma 4.4 and Proposition 4.5, now yield that $\underline{N}$ is an α-critical R-module, hence uniform and therefore $\underline{M}$ is uniform too. □

4.7. Theorem. Let R be a graded ring, $M \in R\text{-gr}$ a uniform graded module. Then $\underline{M}$ is a uniform R-module.

Proof. For $p \in \mathbb{Z}$ denote $\bigoplus_{i \geq p} M_i$ by $M \geq p$. Let $X \subset \underline{M}$ be a nonzero submodule such that $X \cap M_{\geq p} = o$ for some $p \in \mathbb{Z}$. Then $X_{\sim} \cap M_{\geq p} = o$; indeed if x is homogeneous in $X_{\sim} \cap M_{\geq p}$ then there is a $y \in X$, $y = y_1 + \dots + y_m$, with $y_1 = x$ and $\deg y_1 < \dots < \deg y_m$. Since $\deg y_1 \geq p$, $y \in M_{\geq p}$, hence $y \in X \cap M_{\geq p} = o$. So we have $(X_{\sim})_i = o$ for all $i \geq p$. Because $X_{\sim} \neq o$ and $X_{\sim}$ being uniform in R-gr, Lemma 4.6 may be used to derive that $X_{\sim}$ is a uniform R-module, hence $\underline{M}$ is a uniform R-module. Let us suppose now that $X \cap M_{\geq p} \neq o$ for any nonzero submodule X of $\underline{M}$ and for all $p \in \mathbb{Z}$. In this case we proceed to show that M^+ is a uniform R^+-module. By Lemma 4.6 it will be sufficient to establish uniformity of M^+ in R^+-gr. Let $x,y \in h(M^+)$ be nonzero and put $n = \deg x$, $m = \deg y$ with $m \leqslant n$. Since M is uniform $Rx \cap Ry \neq o$ thus $Rx \cap Ry \cap M_{\geqslant n} \neq o$ and we may pick a nonzero homogeneous z in this set. We have $z = \lambda x = \mu y$ with $\lambda, \mu \in h(R)$ and by the choice of z it also follows that $\lambda, \mu \in R^+$, hence $z \in R^+x \cap R^+y$. Consequently M^+ is uniform in R^+-gr. Further, if X,Y are nonzero submodules of $\underline{M}$ then $X \cap M^+ \neq o$ and $Y \cap M^+ \neq o$ yields $X \cap Y \cap M^+ \neq o$ hence $X \cap Y \neq o$.

4.8. Corollary. Let R be a graded ring which is left Noetherian. Let $M \in R$-gr. The Goldie dimension of M in R-gr is equal to the Goldie dimension of the R-module $\underline{M}$.

Proof. From the above theorem and Lemma I.3.3.13. □

Lemma 4.9. Let M be left Noetherian in R-gr, then M has a composition series $M \supset M_1 \supset \ldots \supset M_n = o$, where M_{i-1}/M_i is a critical module for each $1 \leqslant i \leqslant n$.

Proof. As in the ungraded case, cf. [12]. □

Lemma 4.10. Let $M \in R$-gr have Krull dimension, $\text{K.dim}_{R\text{-gr}} M = \alpha$. Then $\text{K.dim}_{R_0} M_i \leqslant \alpha$ for any $i \in \mathbb{Z}$.

Proof. For any graded submodule N_i of M_i we have $N_i = RN_i \cap M_i$, then apply Lemma I.5.4. □

Lemma 4.11. Let M be an α-critical object in R-gr. Then $\text{K.dim}_{R^+} M^+ \leqslant \alpha+1$ and $\text{K.dim}_{R^-} M^- \leqslant \alpha+1$.

Proof. Let $x \in h(M^+)$ be nonzero. We shall show by transfinite induction on α that $\text{K.dim}_{R^+} M^+/R^+x < \alpha$. For $\alpha = o$, M is a simple object in R-gr. From the structure of simple objects in R-gr (see Theorem 6.3 in this chapter) the assertion follows. In case $\alpha \neq o$, the fact that $Rx \neq o$ implies that $\text{K.dim}_{R\text{-gr}} M/Rx < \alpha$ and from the induction hypothesis combined with Lemma 4.1, Lemma 4.9. it results that $\text{K.dim}_{R^+} (M/Rx)^+ \leqslant \alpha$. Obviously : $(M/Rx)^+ = M^+/(Rx)^+$ and $R^+x \subset (Rx)^+$. So we have the exact sequence :

$$o \longrightarrow (Rx)^+/R^+x \longrightarrow M^+/R^+x \longrightarrow M^+/(Rx)^+ \longrightarrow o$$

If $\deg x = k$, then $(Rx)^+/R^+x \cong R_{-k}x \oplus \dots \oplus R_{-1}x$ (isomorphism of R_0-modules). By Lemma 4.10 we obtain $K.\dim_{R^+} (Rx)^+/R^+x \leq \alpha$. Finally Proposition 4.3 applied to the above exact sequence yields : $K.\dim_{R^+} M^+ \leq \alpha+1$. In the similar way it is proven that $K.\dim_{R^-} M^- \leq \alpha+1$.

4.12. Corollary. Let $M \in R\text{-gr}$ be left Noetherian with $K.\dim_{R\text{-gr}} M = \alpha$. Then : $K.\dim_{R^+} M^+ \leq \alpha+1$, $K.\dim_{R^-} M^- \leq \alpha+1$.

Proof. Directly from 4.11, 4.9 and 4.1. □

4.13. Theorem. Let M be left Noetherian in R-gr with $K.\dim_{R\text{-gr}} M = \alpha$, then : $\alpha \leq K.\dim_R \underline{M} \leq \alpha+1$.

Proof. By Lemma I.5.4 : $\alpha \leq K.\dim_R \underline{M}$. By Lemma I.5.5 and Proposition 1.2 we obtain:

$$K.\dim_R \underline{M} \leq \sup(K.\dim_{R\text{-gr}} M,\ K.\dim_{R^-} M^-)$$

So Corollary 4.12 finishes the proof. □

4.14. Corollary. Let $M \in R\text{-gr}$ be left Noetherian and α-critical. Suppose $K.\dim_R \underline{M} = \alpha+1$, then :

1. $\underline{M}$ is an $(\alpha+1)$-critical R-module.
2. $M^+ (M^-)$ is an $(\alpha+1)$-critical R^+-$(R^-$-$)$ module.

Proof. Let $X \subset \underline{M}$ be a nonzero R-submodule. Obviously : $K.\dim_{R^+} M^+ = K.\dim_{R^-} M^- = \alpha+1$. If $X \cap M_{\leq p} = o$ then $X^\sim \cap M_{\leq p} = o$. Considering that : $K.\dim_{R\text{-gr}} M(n) = \alpha$ for all $n \in \mathbb{Z}$, we may assume that $X \cap M^- \neq o$. Then $K.\dim_{R\text{-gr}} M/X^\sim < \alpha$ and $K.\dim_{R^-} M^- \leq \alpha$ (see the proof of Lemma 4.11). By Proposition 1.2 and Corollary 1.3 we obtain : $K.\dim_R \underline{M} \leq \alpha$, hence $\underline{M}$ is $(\alpha+1)$-critical. The second assertion results from foregoing results. □

If $M \in R\text{-gr}$ is an α-critical left Noetherian object, then one expects that if $K.\dim_R M = \alpha$, then $\underline{M}$ is α-critical. For the commutative case this is indeed the case, it follows from :

4.15. Proposition. Let R be a commutative graded ring and let $M \in R\text{-gr}$ be α-critical

Then $\underline{M}$ is an α-critical or an $(\alpha+1)$-critical R-module.

Proof. It is clear that M is α-critical in R-gr if and only if there exists a graded ideal I such that for any $x \in h(M)$, $x \neq o$ we have $\mathrm{Ann}\, x = I$ and $\mathrm{K.dim}_{R\text{-}gr} R/I = \alpha$. Let $x \in M$ be a nonzero element with $x = x_1 + \dots + x_n$ with $\deg x_1 < \dots < \deg x_n$. It is clear that $I \subset \mathrm{Ann}\, x$. If $\lambda \in \mathrm{Ann}\, x, \lambda = \lambda_1 + \dots + \lambda_k$, with $\deg \lambda_1 < \dots < \deg \lambda_k$, then $\lambda x = o$ yields $\lambda_1 x_1 = o$ i.e. $\lambda_1 \in I$. Then $\lambda_1 x_2 = \dots = \lambda_1 x_n = o$, hence $\lambda_1 \in \mathrm{Ann}\, x$. Further, we obtain $\lambda_2, \dots, \lambda_k \in I$, hence $\mathrm{Ann}\, x = I$. Since R/I is a domain, R/I is α-critical or $(\alpha+1)$-critical as an R-module, i.e., M is an α-critical or $(\alpha+1)$-critical R-module. □

4.16. Proposition. Let R be a graded ring, $M \in$ R-gr. M has Krull dimension if and only if for each k, $o \leq k < d$, $M^{(d,k)}$ has Krull dimension over $R^{(d)}$. In this case,

$$\mathrm{K.dim}_R\, M = \sup_{o \leq k \leq d} \mathrm{K.dim}_{R^{(d)}}\, M^{(d,k)} .$$

Proof. From Remark 3.10 it follows that $\mathrm{K.dim}_{R^{(d)}} M^{(d,k)} \leq \mathrm{K.dim}_R\, M$ for all $o \leq k \leq d-1$, hence : $\sup_{o \leq k < d} \mathrm{K.dim}_{R^{(d)}} M^{(d,k)} \leq \mathrm{K.dim}_R\, M$. On the other hand, the lattice of graded submodules of M maps into the product of lattices of graded submodules of $M^{(d,k)}$ with $o \leq k < d$, as follows :

$$N \longrightarrow (N^{(d,o)}, N^{(d,1)}, \dots, N^{(d,d-1)}) .$$

This mapping is strictly increasing so we may apply lemma I.5.5. □

4.17. Corollary. Let K be a graded ring having Krull dimension. For every $d \geq 1$, $R^{(d)}$ has Krull dimension and

$$\mathrm{K.dim}_{R\text{-}gr}\, R = \mathrm{K.dim}_{R\text{-}gr}\, R^{(d)} .$$

In the sequel of this section let R be a commutative graded ring. Denote by $\mathrm{Spec}_g R$ (Spec R) the set of graded prime ideals (prime ideals) of R. By transfinite induction we define on $\mathrm{Spec}_g R$ the filtration : $(\mathrm{Spec}_g R)_o = \{$graded maximal ideals$\}$ $(\mathrm{Spec}_g R)_\alpha = \{p \in \mathrm{Spec}_g R,\ p \subsetneq q$ and $q \in \mathrm{Spec}_g R$ implies $q \in \bigcup_{\beta < \alpha} (\mathrm{Spec}_g R)_\beta$.

If R is a Noetherian graded ring, there exists a smallest ordinal α (resp. β) such that $\mathrm{Spec}_g R = (\mathrm{Spec}_g R)_\alpha$ ($\mathrm{Spec}\, R = (\mathrm{Spec}\, R)_\beta$). The ordinal $\alpha,(\beta)$, is denoted by gcl.dim R (resp. cl.dim R) and will be called the classical graded Krull dimension of R (resp. classical Krull dimension of R). It is well-known, cf. [9] p. 425, that $\mathrm{K.dim}_{R\text{-gr}}\, R = \mathrm{gcl.dim}\, R$, $\mathrm{K.dim}\, R = \mathrm{cl.dim}\, R$. From Theorem 4.13 we derive :

4.18. Corollary. If R is a Noetherian graded ring then $\mathrm{gcl.dim}\, R \leq \mathrm{cl.dim}\, R \leq \mathrm{gcl.dim}\, R + 1$.

4.19. Remark. If K is a graded division ring then we have $\mathrm{gcl.dim}\, K = o$, $\mathrm{cl.dim}\, K = 1$. If $R = k[X,Y]$, k a field and $XY = YX = o$, put $R^+ = k[X]$, $R^- = k[Y]$, then $\mathrm{K.dim}_R\, R = \mathrm{K.dim}\, R^+ = \mathrm{K.dim}\, R^- = 1$.

II.5. THE KRULL DIMENSION OF SOME CLASSES OF RINGS.

Here we include explicit formula for the Krull dimension of certain rings, in terms of the data used in the construction.

5.1. Polynomial Rings.

Let R be any ring, φ a ring automorphism $R \to R$, and consider the ring of twisted polynomials $R[X,\varphi]$. If M is a left R-module, then $M[X,\varphi]$ stands for the left $R[X,\varphi]$-module $R[X,\varphi] \otimes_R M$. Then $M[X,\varphi]$ becomes a graded $R[X,\varphi]$-module if we equip it with the grading : $M[X,\varphi]_i = \{X^i \otimes m, m \in M\}$, for $i \geq o$. In fact, $M[X,\varphi]$ may be identified with the module of polynomials $m_0 + Xm_1 + X^2m_2 + \ldots$, $m_i \in M$, with scalar multiplication given by : $aX^p.X^q m = X^{p+q}\, \varphi^{-p-q}(a)m$, $a \in R$, $m \in M$.

5.1.1. Theorem. 1. $M[X,\varphi]$ has Krull dimension if and only if M is left Noetherian, in this case :

$$\mathrm{K.dim}_{R[X,\varphi]}\, M[X,\varphi] = 1 + \mathrm{K.dim}_R\, M.$$

2. If M is left Noetherian and α-critical then $M[X,\varphi]$ is $(\alpha+1)$-critical.

Proof. Put $S = R[X,\varphi]$, $N = M[X,\varphi]$. If M is left Noetherian then N is left Noetherian in S-mod, hence its Krull dimension is defined. Conversely, assume that N has Krull dimension and that M is not left Noetherian, i.e. there exists a strictly increasing sequence of submodules of M :

$$M_0 \subsetneq M_1 \subsetneq M_2 \subset \dots \subsetneq M_n \subsetneq \dots$$

Set :

$$A = 1 \otimes M_1 + X \otimes M_2 + \dots + X^k \otimes M_{k+1} + \dots$$

$$B = 1 \otimes M_0 + X \otimes M_1 + \dots + X^k \otimes M_k + \dots$$

Clearly $B \subset A$ are graded submodules of N and $XA \subset B$. Let π_i be the canonical map $M_i \to M_i/M_{i-1}$, $i = 1,2,\dots$. Note that M_i/M_{i-1}, $i = 1,\dots,$ is an S-module with scalar multiplication :

$$(a_0 + a_1 X + \dots).\pi_i(m) = \pi_i(\varphi^{1-i}(a_0)m) \text{ for every } m \in M_i \ .$$

Define the mapping $\theta : A/B \to M_1/M_0 \oplus M_2/M_1 \oplus \dots$ as follows, if $f = 1 \otimes m_1 + X \otimes m_2 + \dots \in A$, pu $\theta(f \bmod B) = (\pi_1(m_1), \pi_2(m_2), \dots)$. It is straightforward to check that θ is an S-isomorphism. Since N has Krull dimension, so does A/B, cf. Lemma 4.1., and therefore it is of finite Goldie dimension, contradiction.

Because of Lemma 4.9. we may assume that M is α-critical. Proposition 4.5 and the fact that N is a graded S-module yield that we only have to prove that N is $(\alpha+1)$-critical in S-gr then the equality in 1 and also 2 will follow. Consider the following infinite, strictly decreasing sequence :

$$N \supsetneq XN \supsetneq X^2N \supsetneq \dots \supsetneq X^iN \supsetneq \dots,$$

where $X^iN/X^{i+1}N \cong N/XN \cong M$.

Obviously : $K.\dim_S N \geq \alpha+1$.

Let $z = X^k \otimes m$ be a nonzero element of degree k in h(N) and consider the following sequence of graded modules :

$$(\star) \quad N \supset XN \supset \dots \supset X^kE \supset Sz \ ,$$

where $K.\dim_S X^iN/X^{i+1}N = \alpha$ and $K.\dim_S X^kN/Sz = Rm[X,\varphi]$. Since M is α-critical it follows that $K.\dim_R (M/Rm)[X,\varphi] \leq \alpha$ Lemma 4.1 applied to (*) yields that $K.\dim_S N/Sz \leq \alpha$ and therefore N is $(\alpha+1)$-critical. □

5.1.2. Corollary. In case we also consider a φ-derivation δ then we have :
If $M \in R$-mod is left Noetherian then

$$K.\dim_R M \leq K.\dim_{R[X,\varphi,\delta]} M[X,\varphi,\delta] \leq 1 + K.\dim_R M ,$$

where $M[X,\varphi,\delta] = R[X,\varphi,\delta] \otimes_R M$.

Proof. $R[X,\varphi,\delta]$ is a filtered ring with associated graded ring equal to $R[X,\varphi]$ whereas $M[X,\varphi,\delta]$ is a filtered module with associated graded module $M[X,\varphi]$. The foregoing theorem and Proposition I.5.8 finish the proof. □

5.2. Rings of Formal Power Series.

Let $R[X,\varphi]$ be as in 5.1. If $M \in R$-mod then $M[[X,\varphi]]$ is the "module of formal power series" consisting of elements $m_0 + xm_1 + x^2m_2 + \dots$, $m_i \in M$ and with scalar multiplication given by : $(aX^p).(X^qm) = X^{p+q}\varphi^{-p-q}(a)m$, $a \in R$, $m \in M$. $R[[X,\varphi]]$ is the ring of formal power series and $M[[X,\varphi]]$ is a left $R[[X,\varphi]]$-module, generally not isomorphic to $R[[X,\varphi]] \otimes_R M$!

5.2.1. Theorem. 1. $M[[X,\varphi]]$ has Krull dimension if and only if M is left Noetherian and in that case :

$$K.\dim_{R[[X,\varphi]]} M[[X,\varphi]] = 1 + K.\dim_R M .$$

2. If M is left Noetherian and α-critical, then $M[[X,\varphi]]$ is $(\alpha+1)$-critical.

Proof. The module $M[[X,\varphi]]$ endowed with the X-adic topology is complete and its associated graded module is $M[X,\varphi]$. The proof for the first part of statement 1. is only a slight modification of the proof of Theorem 5.1.1. (1), while the equality in 1. and statement 2. are direct consequences of Theorem 5.1.1. and Corollary I.5.6. □

5.2.2. Corollary. If R is left Noetherian and $\varphi : R \to R$ a ring automorphism, then :

$$\text{K.dim } R[[X,\varphi]] = 1+\text{K.dim } R.$$

5.3. The enveloping Algebra of a Lie Algebra.

5.3.1. Theorem. Let K be a commutative field, g a finite dimensional Lie K-algebra. The enveloping algebra of g is a left and right Noetherian ring and its Krull dimension is at most $[g : k] = n$.

Proof. Let U be the enveloping algebra of g and let U_n be the K-subspace of U generated by 1 and products of the form $g_1, g_2, \ldots, g_m$ with $m \leq n$, $g_i \in g$. In this way U is a filtered ring which is discrete, and its associated graded ring is the ring of polynomials $K[X_1,\ldots,X_n]$. So application of Proposition I.5.8 finishes the proof. □

5.4. Weyl Algebras.

Let k be an (algebraically closed) field of characteristic zero. Let $A_1(k)$ be the k-algebra generated by the elements p and q satisfying $[p,q] = pq-qp = 1$; $A_n(k)$ will be the k-algebra generated by the 2n elements $\{p_i, q_j, 1 \leq i,j \leq n\}$ subjected to the relations $[p_i,q_i] = 1$ $[p_i,q_j] = [p_i,p_j] = [q_i,q_j] = o$ if $i \neq j$. So $A_n(k) = A_{n-1}(k) \otimes_k A_1(k)$ Let us also define $A_o(k) = k$. Now $A_n(k)$ may be characterized as the algebra generated over $A_{n-1}(k)$ by elements p,q commuting with $A_{n-1}(k)$ and satisfying $qp-pq = 1$. Therefore $x \in A_n(k)$ may be written as $\Sigma\, x_{\alpha\beta}\, p^\alpha q^\beta$, with $x_{\alpha\beta} \in A_{n-1}(k)$. By means of the total degree in p and q we may equip $A_n(k)$ with a discrete filtration such that the associated graded ring $G(A_n(k)) \cong A_{n-1}[X,Y]$. Consequently $A_n(k)$ is a left and right Noetherian domain.

5.4.1. Proposition. 1. $A_1(k)$ is simple, i.e. there are no nonzero proper ideals. 2. For every nonzero left ideal L of A_1, A_1/L is of finite length. In particular $\text{K.dim } A_1 = 1$.

Proof. 1. The following relations are easily verified : $[p,p^\alpha q^\beta] = \beta p^\alpha q^{\beta-1}$ and $[q,p^\alpha q^\beta] = -\alpha p^{\alpha-1} q^\beta$. Let I be a nonzero ideal of A_1, $u \neq o$, $u \in I$, say $u = \Sigma u_{\alpha\beta} p^\alpha q^\beta$. Since $[p,u]$ and $[q,u] \in I$, the relations established imply $I \cap k \neq (o)$, therefore $I = A_1$.

2. Let L be a nonzero left ideal of A_1 and consider the strictly decreasing sequence of left ideals containing L :

$$A_1 = L_0 \supset L_1 \supset \dots \supset L_n \supset \dots .$$

Apply the functor G; we get :

$$k[X,Y] \cong G(L_0) \supset G(L_1) \supset \dots \supset G(L_n) \supset \dots .$$

Since $G(L) \neq o$ we have that K.dim $k[X,Y]/G(L) \leq 1$. It follows that, for n sufficiently large, $G(L_n)/G(L_{n+1})$ is a $k[X,Y]/G(L)$-module of finite length. Therefore it has to be finite dimensional over k (indeed, if M is a maximal ideal of $k[X,Y]$ then $k[X,Y]/M$ is an algebraic extension of k hence finite dimensional). So, for large n, L_n/L_{n+1} is a finite dimensional k-space. Because of the density theorem it now follows from the above that A_1/L has to be left Artinian and thus of finite length i.e. K.dim $A_1 = 1$.

□

Note that, since A_1 is a quasi-simple ring, the proof of 5.4.1. 1. also may be extended to prove that A_n is simple for all n.

<u>5.4.2. Lemma.</u> Let R be a ring and let S_1, S_2 be multiplicatively closed sets such that R satisfies the left Ore conditions with respect to S_1 and S_2. Suppose that for any $s_1 \in S_1$, $s_2 \in S_2$ we have that $Rs_1 + Rs_2 = R$, then for left ideals $I \subset J$ of R such that $S_1^{-1}I = S_1^{-1}$ and $S_2^{-1}I = S_2^{-1}J$, we have $I = J$.

<u>Proof.</u> If $M \in R$-mod is such that $S_1^{-1}M = S_2^{-1}M = o$ then for any $x \in M$ there exist $s_1 \in S_2$, $s_2 \in S_2$ such that $s_1 x = s_2 x = o$, hence $(Rs_1 + Rs_2)x = o$ or $x = o$. Applying this to J/I yields the statement. □

<u>5.4.3. Theorem.</u> For any $n \geq 1$, K.dim $A_n(k) = n$.

<u>Proof.</u> Put $S_1 = k[p_n] - \{o\}$, $S_2 = k[q_n] - \{o\}$. Obviously A_n satisfies the left and right Ore conditions with respect to these sets. Since $A_n \cong (A_{n-1}[X])[Y,1,\delta]$, where δ is the derivation of $A_{n-1}[X]$ given by common derivation in X : $\frac{\partial}{\partial x}$, it follows easily

that : $S_1^{-1} A_n \cong A_{n-1}(k(X))[Y,1,\delta]$.

Hence, the induction hypothesis on n and Theorem 5.1.1. imply that K.dim $S_1^{-1} A_n \leq n$. Similarly one establishes K.dim $S_2^{-1} A_n(k) \leq n$. Put $s_1 = P(p_n) \in S_1$, $s_2 = Q(q_n) \in S_2$, $L = A_n^{(k)} s_1 + A_n^{(k)} s_2$. Clearly $L \cap k[p_n] \neq o$, $L \cap k[q_n] \neq o$, so it follows that $L = A_n(k)$ and we may apply the lemma to get that K.dim $A_n(k) \leq n$. It remains to establish $n \leq$ K.dim $A_n(k)$. Consider the $A_1(k)$-module $A_1(k)/A_1(k)q_n$ which is simple and has commutator exactly k. It is known that the $A_{n-1}(k) \otimes A_1$-submodules of $A_{n-1}(k) \otimes A_1/A_1(k)q$ may be represented as $I \otimes A_1(k)/A_1(k)q$ where I is some left ideal of A_{n-1}. Then K.dim $A_{n-1}^{(k)} = \text{K.dim}_{A_1} (A_{n-1}^{(k)} \otimes A_1(k)/A_1(k)q$. Now consider the sequence :

$$A_n(k) = A_{n-1}(k) \otimes A_1(k) \supset A_{n-1}(k) \otimes A_1(k)q \supset A_{n-1}(k) \otimes A_1 q^2 \supset \ldots$$

The Krull dimension of each factor is K.dim $A_{n-1}(k)$, hence K.dim $A_m(k) \geq 1 +$ K.dim $A_{n-1}(k)$ i.e. K.dim $A_n(k) \geq n$. □

II.6. GRADED DIVISION RINGS.

6.1. Lemma. If R is a graded domain then left-invertible elements are homogeneous.

Proof. Let $x = \sum_{i=n}^{N} x_i \in R$, suppose $\sum_{j=m}^{M} y_j \sum_{i=n}^{N} x_i = 1$ then $y_M x_N \neq o$ and $y_m x_n \neq o$ yields $M + N = o$, $m + n = o$ but then $M \geq m$, $N \geq m$ yields $N = n$, $M = m$. □

Note that if R is not a domain, the lemma does not hold anymore, as may be seen in case $R = k[X]/(X)^n$, where 1-X is invertible. If every nonzero homogeneous element of a graded ring R is invertible then R is said to be a graded division ring. Note that this definition implies that a graded division ring R is a domain and that R_0 is a field.

6.2. Lemma. Let R be a graded ring. The following properties of R are equivalent:

1. R has no nonzero proper graded left ideals.
2. R has no nonzero proper graded right ideals.
3. R is a graded division ring.

Proof. Straightforward.

6.3. Theorem. If R is a graded division ring then $R = R_0$ or $R \cong R_0[X, X^{-1}, \varphi]$ where $\varphi : R_0 \to R_0$ is an automorphism, X an indeterminate of degree $t > 0$ such that $Xa = \varphi(a)X$ and $X^{-1}a = \varphi^{-1}(a)X^{-1}$.

Proof. If $R \neq R_0$ then there exists a homogeneous element a with smallest positive degree, t say. Since $R = Ra$, $R_t = R_0 a$, and for all $i \in \mathbb{Z}$, $i \notin (t)$, $R_i = 0$; moreover $a^n \neq 0$ yields $R_{nt} = R_0 a^n$ for all $n \in \mathbb{Z}$. Let $\lambda \in R_0$ be arbitrary, then there is a unique element $\varphi(a) \in R_0$ such that $a\lambda = \varphi(\lambda)a$. It is easily seen that $\varphi : R_0 \to R_0$ is a field isomorphism and we have equalities $a^n\lambda = \varphi^n(\lambda)a^n$, $a^{-1}\lambda = \varphi^{-1}(\lambda)a^{-1}$. The graded ring homomorphism $\alpha : R_0[X, X^{-1}, \varphi] \longrightarrow R$ defined by $\alpha(X) = a$, $\alpha(X^{-1}) = a^{-1}$ is an isomorphism (note that we put $\deg X = t$). □

6.4. Corollary. Graded division rings are fields or two-sided principal ideal rings.

Proof. If the graded division ring R is not a field then it is isomorphic to the ring of fractions with respect to the multiplicative system $\{1, X, X^2, \ldots\}$ of $R_0[X, \varphi]$. The latter is left- and right-principal because of our results in Section II.2., hence R is left- and right-principal too. □

II.7. THE STRUCTURE OF SIMPLE OBJECTS IN R-gr.

7.1. Definition. Let R be a graded ring. An $S \in R$-gr is said to be simple if 0 and S are its only graded submodules.

7.2. Lemma. Let $S \in R$-gr be simple, then for each $i \in \mathbb{Z}$, $S_i = 0$ or S_i is a simple R_0-module.

Proof. Suppose $S_i \neq 0$. If $x \neq 0$, $x \in S_i$ then $Rx = S$, hence $R_0 x = S_i$. Consequently S_i is a simple R_0-module. □

7.3. Remarks. 1. If R is positively graded and $S \in R$-gr is simple then there is a $j \in \mathbb{Z}$ such that $S = S_j$. Indeed if S_i and S_j are both nonzero, select x_i, $x_j \neq 0$ in S_i, S_j

resp, then $x_i = rx_j$, $x_j = r'xi$ for some $r, r' \in h(R)$ and depending whether $i > j$ or $j > i$, one of these relations is impossible, hence $i = j$.

2. If S is simple in R-gr and S is left-or right-limited, then $\underline{S}$ is a simple R-module. This follows directly from Corollary 1.3.

<u>7.4. Lemma</u>. Let R be a graded ring. S a simple object of R-gr. Then $D = \mathrm{END}_R(S)$ is a graded division ring and if $D_o \neq D$ then $\underline{S}$ is a 1-critical R-module.

<u>Proof</u>. Let $f : S \to S$ be a nonzero graded morphism of degree n. Then f may be represented as a morphism of degree $o : S \to S(n)$ but as S and S(n) are simple objects in R-gr, f has to be an isomorphism in R-gr, i.e. invertible.

To prove the second statement we first prove that for any nonzero R-submodule X of $\underline{S}$, $\underline{S}/X$ has finite length. Since $X \neq o$, $X_{\sim} \neq o$ hence $X_{\sim} = S$. We may assume $X \cap S^+ \neq o$, because if $X \cap S^+ = o$ then $X_{\sim} \cap S^+ = o$ hence $S^+ = o$ i.e. S is right-limited and $\underline{S}$ is simple in R-mod by Remark 7.3.2. Now let $X_1 \supset X_2 \supset \dots \supset X_n \supset \dots$, be a descending chain of submodules of S containing X. Then we obtain the descending chain of R^+-submodules in S^+ :

$$X_1 \cap S^+ \supset X_2 \cap S^+ \supset \dots \supset X_n \cap S^+ \supset \dots ,$$

and also the descending chain of graded submodules of S^+ :

$$(X_1 \cap S^+)^{\sim} \supset (X_2 \cap S^+)^{\sim} \supset \dots \supset (X_n \cap S^+)^{\sim} \supset \dots \supset (X \cap S^+)^{\sim}$$

where the operation $\sim$ is defined in S^+. Now the only graded R^+-submodules of S^+ are $S_{\geqslant p}$ with $p \geqslant o$, (the proof of this claim is part 1. of the following Theorem 7.5) thus we may write : $(X \cap S^+)^{\sim} = S_{\geqslant p}$ and $(X_k \cap S^+)^{\sim} = S_{\geqslant p_j}$, where $p_1 \leqslant p_2 \leqslant \dots \leqslant p_j \leqslant \dots \leqslant p$. It follows that for some $j : p_j = p_{j+1} = \dots$, whence $(X_j \cap S^+)^{\sim} = (X_{j+1} \cap S^+)^{\sim} = \dots$.

By Proposition 1.2 and Corollary 1.3 we obtain that $X_j = X_{j+1} = \dots$. Therefore $\underline{S}/X$ is an Artinian object of R-mod, whereas by Theorem 3.3. $\underline{S}$ is a left Noetherian R-module, so $\underline{S}/X$ has finite length.

Supposing that D has nontrivial grading, then we have $D \cong D_o[X, X^{-1}, \varphi]$ because of Theorem 6.3. The endomorphism of S represented by 1-X is an injective morphism which is not invertible. We obtain a strictly descending chain of R-submodules of $\underline{S}$: $\underline{S} \supset (1-X)\underline{S} \supset (1-X)^2\underline{S} \supset \dots$, hence $\underline{S}$ is not left Artinian. Therefore $K.\dim_{\underline{R}}\underline{S} = 1$ and the above implies that $\underline{S}$ is 1-critical. □

7.5. Theorem. Let S be simple in R-gr, then :

1. The only graded R^+-submodules of S^+ (resp. R-submodules of S^-) are $S_{\geq p}$ (resp. $S_{\geq -p}$) for $p \geq o$.
2. Any R^+-submodule of S^+ and R^--submodule of S^- is principal.
3. Any R-submodule of $\underline{S}$ is principal.
4. $\underline{S}$ is either simple in R-mod or a 1-critical R-module.
5. The intersection of all maximal R-submodules of $\underline{S}$ is zero.

Proof. 1. Suppose that $M = \bigoplus_{i \geq o} M_i$ is a graded submodule of S^+. Let p be the smallest natural number such that $M_p \neq o$, then, by Lemma 7.2., we have that $M_p = S_p$. Further, the fact that $RS_p = S$ then yields $R^+S_p = \bigoplus_{i \geq p} S_i = S_{\geq p}$ and thus $M \supset R^+M_p = S_{\geq p}$ follows.

2. Easy.

3. Let $X \subset \underline{S}$ be a nonzero R-submodule. If $X \cap S^+ = o$, then $(X \cap S^+)_\sim = o$ but also $X_\sim \cap S^+ = o$. Indeed, $x \in h(X_\sim \cap S^+)$ and $x \neq o$ means that there is a nonzero $y \in X$ such that $y = y_1 + \dots + y_m$ with $y_1 = x$ and $\deg y_1 < \dots < \deg y_m$. Since $\deg x \geq o$, $y \in S^+$ and hence $y \in X \cap S^+$, contradiction. $X_\sim$ is not S since $X_\sim \cap S^+ = o$ but this contradicts simplicity of S, hence we may assume that $X \cap S^+ \neq o$ and then we have that $(X \cap S^+)^\sim = S_{\geq p}$ for some $p \in \mathbb{N}$ ($\sim$ operates in S^+). Pick $x_p \neq o$ in S_p. There is a $y \in X \cap S^+$ such that $y = y_1 + \dots + y_m$ with $y_m = x_p$ and $\deg y_1 < \dots < \deg y_m$. But this means that :

$$(X \cap S^+)^\sim \supset (R^+y)^\sim \supset Rx_p = S_{\geq p} ,$$

hence $(X \cap S^+)^\sim = (R^+y)^\sim$ and therefore $X \cap S^+ = R^+y$. On the other hand $X \cap S^+ \supset Ry \cap S^+ \supset R^+y$, whence $X \cap S^+ = Ry \cap S^+$ follows. Because $Ry \neq o$, $(Ry)_\sim = S = X_\sim$ and then Corollary 1.3 entails that $X = Ry$. □

4 & 5. To prove these statements we need some preliminary facts, which we will present here as sublemmas.

Let M be a graded R-module. M is a left graded module over $S = END_R(M) = HOM_R(M,M)$ and the graded ring $B^g(M) = END_S(M)$ is called the <u>graded bi-endomorphism ring</u> of M. The canonical morphism $\rho : R \to B^g(M)$ defined by $\rho(r)(x) = rx$, is a morphism of graded rings.

<u>Sublemma 1</u>. (The Density Theorem).

Let M be a semi-simple object in R-gr (i.e. direct sum of simple objects), let $x_1, \dots, x_n \in h(M)$ and $\alpha \in B^g(M)$ homogeneous. There exists $r \in h(R)$ such that $\alpha(x_i) = rx_i$, $1 \leq i \leq n$.

<u>Sublemma 2</u>. Let L be a minimal graded left ideal of R (i.e. L is simple in R-gr). Then either $L^2 = o$ or $L = Re$ where e is an idempotent homogeneous element of R. Moreover the ring eRe is a graded division ring. Conversely if R is a semi-prime ring, e an idempotent homogeneous element of R such that eRe is a graded division ring then $L = Re$ is minimal graded in R. If <u>eRe</u> is a field (i.e. eRe is trivially graded) then I is a minimal left ideal of R.

Both sublemmas may be proved in a way similar to the method of proof in the ungraded case. Now let us return to the proof of the theorem.

4. Since $\underline{S}$ is a finitely generated R-module we have $D = END_R S = End_R \underline{S}$ which is a graded division ring. If $\underline{D}$ is not a field then Lemma 7.4. yields that $\underline{S}$ is 1-critical. Suppose $D = D_o$ is a field and let P be the annihilator in R of S. Clearly P is a graded prime ideal and since S is an R/P-module we may reduce the proof to the case where $P=o$.

According to sublemma 1, R is dense in $B^g(S) = T$. Obviously S is simple in T-gr and S is isomorphic to a graded left ideal L of T. Since $L^2 = L$ and $END_T(L) \cong END_T(S) = D$, the second sublemma may be applied, yielding minimality of $\underline{L}$ in T, hence $\underline{S}$ is a simple T-module. Choose $x \neq o$, $x \in S$ and $y \in S$; put $x = x_1 + \dots + x_n$ with $\deg x_1 < \dots < \deg x_n$. Now $\underline{S}$ being simple in T-mod, there is an $\alpha \in T$ such that $y = \alpha x$. Let $\alpha_1, \dots, \alpha_m$ be the homogeneous components of α. Sublemma 1 provides the existence of homogeneous $r_i \in R$ such that $\alpha_i x_j = r_i x_j$, $1 \leq j \leq n$. Then $\alpha_i x = r_i x$ and hence $y = \alpha x = (\sum_{i=1}^{m} r_i)x$, proving that $\underline{S}$ is a simple R-module.

5. By 4. it suffices to consider the case where $D = END_R(S)$ is a graded division ring with nontrivial grading. Using sublemma 1 again, it results that any maximal T-submodule of $\underline{S}$ is a maximal R-submodule, therefore it will be sufficient to show that the Jacobson radical $J(\underline{S})$ of $\underline{S}$ over the ring T is zero. Now $\underline{S}$ is a projective T-module of finite type and it is well known that in this case :

$$END_R(\underline{S}/J(\underline{S})) \cong END_T(S)/J(END_T(S)) = D/J(D),$$

where $J(D)$ is the Jacobson radical of the ring D. It is easy to calculate that for any graded division ring D, $J(D) = o$. So, if $J(\underline{S})$ were nonzero then $\underline{S}/J(\underline{S})$ is a T-module of finite length and together with $END_T(\underline{S}/J(\underline{S})) \cong D$ this states that D is a field i.e. trivially graded, contradiction! □

For modules of polynomials over simple objects we have :

<u>7.6. Theorem.</u> Let R be an arbitrary ring, S a simple left R-module. Consider the left module $S[X]$ over the ring of polynomials $R[X]$; then we have the following properties :

1. Any $R[X]$-submodule of $S[X]$ is principal.
2. $S[X]$ is 1-critical.
3. The Jacobson radical of $S[X]$ (over the ring $R[X]$) is zero.

<u>Proof.</u> 1. Consider $S[X]$ and $R[X]$ as graded objects with the obvious grading. The graded submodules of $S[X]$ are of the form $\bigoplus_{i \geq p} SX^i$, hence principal. By Proposition I.5.8 (3), statement 1 follows.

2. It is easy to show that $S[X]$ is 1-critical in $R[X]$-gr. From Proposition 4.5 we then deduce that $S[X]$ is 1-critical in $R[X]$-mod.

3. Put $D = END_R(S)$, $B^g(S) = END_D(S)$. The ring $B^g(S)$ is a regular ring in the sense of von Neumann. As a $B^g(S)$ module S is isomorphic to a minimal left ideal L of $B^g(S)$. The canonical morphism $\rho : R \to B^g(S)$, $\rho(r) = \varphi_r$ where $\varphi_r(x) = rx$, extends to a morphism $R[X] \to B^g(S)[X]$. Now $S[X]$ is $B^g(S)[X]$-module isomorphic to $L[X]$, which is a direct summand in $B^g(S)[X]$. Furthermore, $B^g(S)$ being regular, a classical result of Amitsur

yields that the Jacobson radical of $B^g(S)[X]$ is zero, hence the Jacobson radical of $R[X]$ is zero. The proof of 3 may now be finished by showing that maximal $B^g(S)[X]$-submodules M of $S[X]$ are also maximal $R[X]$-submodules. Pick $s(X) = s_0 + s_1 X + \dots + s_j X^j \in S[X] - M$. Then we have : $M + B^g(S).s(X) = S[X]$. Let $b_0 + b_1 X + \dots + b_m X^m \in B^g(S)[X]$, write $(b_0 + b_1 X + \dots + b_m X^m)(s_0 + s_1 X + \dots + s_j X^j) = s_0 b_0 + (b_0 s_1 + b_1 s_0)X + (b_0 s_2 + b_1 s_1 + b_2 s_0)X^2 + \dots + b_m s_j X^{m+j}$. Because of the density theorem there exists $a_0, a_1, \dots, a_m \in R$, such that $a_0 s_i = b_0 s_i$, $a_1 s_i = b_1 s_i, \dots, a_m s_i = b_m s_i$ for each $o \leq i \leq j$. Thus $B^b(S).s(X) = R[X].s(X)$, hence $M + R[X].s(X) = S[X]$, proving that M is a maximal $R[X]$-submodule of $S[X]$. □

<u>7.7. Corollary</u>. If $M \in R$-mod has finite length n then every $R[X]$-submodule of $M[X]$ may be generated by less than n elements.

<u>Proof</u> : We proceed by induction on n. If $n = 1$ then M is simple and the assertion follows from the foregoing theorem. Suppose now that the statement is true for n and let M have length n+1. Consider a maximal left-submodule M_1 of M. From the exact sequence in R-mod :

$$o \to M_1 \to M \to M/M_1 \to o$$

we obtain an exact sequence in $R[X]$-mod :

$$o \to M_1[X] \to M[X] \to M/M_1[X] \to o .$$

If N is an $R[X]$-submodule of $M[X]$ then $N \cap M_1[X]$ is generated by less than n elements, whereas $N/N \cap M_1[X]$ is a submodule of $M/M_1[X]$ and may therefore be generated by a single element. Consequently N may be generated by less than n+1 elements. □

Let D be a graded division ring, then we denote by d(D) the minimal positive i such that $D_i \neq o$ or $d(D) = o$ if $D_i = o$ for each $i \neq o$. Obviously D has trivial grading if and only if $d = o$.

<u>7.8. Proposition</u>. Let R be a graded ring, $S \in R$-gr a simple object. Then :

1. <u>S</u> is a semisimple R_0-module.

2. Put $D = END_R(S)$, then the number of isotopic components of the R_o-module $\underline{S}$ is less than or equal to d.

Proof : 1. follows from Lemma 7.2.

2. Let $\alpha \in D_d$, $\alpha \neq o$. The mapping $\alpha : S \to S$ is an R-automorphism of degree d, so for any $n \in \mathbb{Z}$, the mapping $S_n \to S_{d+n}$ sending $x \in S_n$ to $\alpha(x)$ is an R_o-isomorphism, whence 2 follows.

7.9. Example. Let K be a trivially graded field and let $V = \underset{n \geq o}{\oplus} V_n$ be a graded K-vector space, where $V_n = K(n) \oplus ... \oplus K(n)$, n+1 terms. Let R be $END_K V$ with $K \subset R_o$. Clearly V is a simple object in R-gr. and as V is a graded R-module which is left limited it results that $\underline{V}$ is a simple R-module. If $n \neq n'$ then $\dim_K V_n \neq \dim_K V_{n'}$ and thus $V_n \not\cong V_{n'}$ in R_o-mod. Consequently as an R_o-module V has infinitely many isotopic components.

II.8. THE JACOBSON RADICAL OF GRADED RINGS.

Let R be a graded ring, $M \in R$-gr. A graded submodule N of M is said to be (graded) maximal in M if M/N is a graded simple R-module. Clearly N is maximal in M if and only if N is proper and for all $x \in h(M)$, $N + Rx = M$. Of course, if N is maximal in M, $\underline{N}$ need not be maximal in $\underline{M}$.

8.1. Lemma. Let S be simple in R-gr. There exists a left graded ideal I of R which is maximal and an integer $n \in \mathbb{Z}$, such that $S \cong (R/I)(n)$.

Proof : Let $x \neq o$, $x \in h(S)$ with $\deg x = m$. Then the morphism $\alpha : R \to S(m)$ defined by $\alpha(r) = rx$ is of degree o. Since S(m) is simple, α is surjective i.e. $S(m) \cong R/I$ with I maximal in R and thus $S \cong (R/I)(-m)$. □

8.2. Definition. If $M \in R$-gr then the graded Jacobson radical of M, denoted by $J^g(M)$ is the intersection of all maximal graded submodules of M, and we define $J^g(M) = M$ if M has no proper maximal graded submodules.

J^g enjoys properties which closely ressemble the corresponding ungraded properties, let us just mention :

8.3. Lemma. Let R be a graded ring, $M \in R\text{-gr.}$, then :

1. If M is of finite type then $J^g(M) \neq M$ or $M = o$.
2. $J^g(M) = \cap\{\text{Ker } f,\ f \in \text{Hom}_{R\text{-gr}}(M,S),\ S \text{ simple object in } R\text{-gr}\}$

   $= \cap\{\text{Ker } f,\ f\ \ \text{HOM}_R(M,S),\ S \text{ simple object in } R\text{-gr}\}$.
3. If $f \in \text{HOM}_R(M,N)$ then $f(J^g(M)) \subset J^g(N)$.
4. $J^g({}_RR) = \cap\{\text{Ann}_R S,\ S \text{ any simple object in } R\text{-gr}\}$.
5. $J^g({}_RR)$ is an ideal of R.
6. $J^g({}_RR)$ is the greatest graded proper ideal satisfying : if $a \in h(R)$ is such that the image of a in $R/J^g({}_RR)$ is invertible then a is invertible in R.
7. $J^g({}_RR) = J^g(R_R)$.

In view of 5. and 7. in Lemma 8.3 we will denote by $J^g(R) = J^g({}_RR) = J^g(R_R)$ the graded Jacobson radical of the ring R.

8.4. Lemma. (Nakayama's lemma).
Let $M \in R\text{-gr}$ be of finite type, then $J^g(R)M \neq M$.

8.5 Corollaries. 1. Let J(R) denote the Jacobson radical of R, then $J(R) \subset J^g(R)$.
2. Let R be positively graded, then $J^g(R) = J(R_o) \oplus R_+$.

Proof : 1. Directly from Theorem 7.5, 5..
2. The ideal $R_+ = \bigoplus_{i>o} R_i$ has the property that $R_+M \neq M$ for every finitely generated graded left R-module M. So if $S \in R\text{-gr}$ is simple, $R_+S = o$ and $R_+ \subset J^g(R)$ follows. Morewe have seen that $S = S_{n_o}$ for some $n_o \in \mathbb{Z}$ and S_{n_o} is a simple R_o-module. Hence $J(R_o)S_{n_o} = o$ and thus $J(R_o) + R_+ \subset J^g(R)$. On the other hand $J^g(R) = J_o \oplus J_{\geq o}$ i.e. $J^g(R) = J_o \oplus R_+$. Consider a simple R_o-module T and consider $S \in R\text{-gr}$ with $S_o = T$, $S_i = o$ if $i \neq o$. Since S is simple in R-gr : $J_oT = J_oS = o$, hence $J_o \subset J(R_o)$. □

8.6. Theorem (Hopkins' theorem).
Let R be a graded ring which is left gr.Artinian, then R is a left Noetherian graded

ring.

Proof : Formally similar to the ungraded case one shows first that $R/J^g(R)$ is a semisimple object of R-gr and that there exists a positive n such that $(J^g(R))^n = o$. Then for any i, $1 \leq i \leq n-1$, $(J^g(R))^i/(J^g(R))^{i+1}$ is annihilated by $J^g(R)$ hence an $R/J^g(R)$-module. It follows that $(J^g(R))^i/(J^g(R))^{i+1}$ is semisimple in R-gr and because it is also left Artinian it has to be of finite type, hence left Noetherian. It is then also obvious that R^R is a Noetherian object in R-gr. □

8.7. Remark. If R is left gr.Artinian then R is not necessarily left Artinian. For example $k[X,X^{-1}]$ with non trivial grading according to the degree in X.

II.9. SEMISIMPLE GRADED RINGS. Goldie's Theorem for Graded Rings.

A graded ring R is said to be semisimple if R-gr is a semisimple category i.e any object of R-gr is semisimple. It is clear that R is semisimple graded if and only if : (★) $R = L_1 \oplus \dots \oplus L_n$, L_i , $1 \leq i \leq n$, being minimal graded left ideals of R. The graded ring R is said to be simple if it has decomposition (★) but with $HOM_R(L_i, L_j) \neq o$ for any $i,j = 1,\dots,n$. The latter condition means that for any couple (i,j) there exists an integer n_{ij} such that $L_j \cong L_i(n_{ij})$ in R-gr. The graded ring R is said to be uniformly simple if R admits the decomposition $R = L_1 \oplus \dots \oplus L_n$ in minimal graded left ideals but with $L_i \cong L_j$ for any i,j.

9.1. Graded Matrix Rings.

If R is a graded ring, $n \in \mathbb{N}$, then $M_n(R)$ denotes the ring of n by n matrices with entries from R. Fix an n-tuple of integers $(d_1,\dots,d_n)$, which will be written as $\bar{d}$. To any $\lambda \in \mathbb{Z}$, associate the following abelian group of matrices.

$$M_n(R)_\lambda(\bar{d}) = \begin{pmatrix} R_\lambda & R_{\lambda+d_2-d_1} & \cdots & R_{\lambda+d_n-d_1} \\ R_{\lambda+d_1-d_2} & R_\lambda & \cdots & \vdots \\ \vdots & \cdots\cdots & \cdots & R_{\lambda+d_n-d_{n-1}} \\ R_{\lambda+d_1-d_n} & \cdots\cdots & \cdots & R_\lambda \end{pmatrix}$$

It is obvious that $M_n(R) = \bigoplus_{\lambda \in \mathbb{Z}} M_n(R)_\lambda(\overline{d})$. In this way we obtain a graded ring $M_n(R)(\overline{d})$; note that $M_n(R)(\overline{o})$ will be denoted by $M_n(R)$ as there will usually be no confusion whether the ring is considered as a graded object or not. With these notations we have :

9.1.1. Lemma. 1. For any $m \in \mathbb{Z}$, $M_n(R)(\overline{d}) = M_n(R)(\overline{d+m})$.

2. $M_n(R)(\overline{d}) \cong M_n(R)(\overline{\sigma(\overline{d})})$ where σ is a permutation of $\overline{d}$.

3. If M is a free graded R-module with finite basis $e_1, \ldots, e_n$ then $\mathrm{END}_R(M) \cong M_n(R)(\overline{d})$ where $d_i = \deg e_i$, $i = 1, \ldots, n$.

4 Let D be a nontrivially graded division ring. Then for any system of integers $d_1', \ldots, d_n'$ there corresponds an n-tupel $d_1, \ldots, d_n$ such that $o \leq d_1 \leq d_2 \leq \ldots \leq d_n < d(D)$ and $M_n(D)(\overline{d'}) \cong M_n(D)(\overline{d})$, where $d(D)$ is positive and minimal such that $D_{d(D)} \neq o$.

9.1.2. Theorem (Wedderburn's Theorem).

Let R be a graded ring. The following statements are equivalent :

1. R is graded simple (resp. uniformly simple).

2. There exists a graded division ring D and $\overline{d} \in \mathbb{Z}^n$ such that $R \cong M_n(D)(\overline{d})$ (resp. $R \cong M_n(D)$).

Proof. As in the ungraded case. □

In connection with this the problem arises to determine the number of isomorphism classes of simple graded rings of seize n over a given graded division ring D. The following proposition will solve this problem. Let us denote $d(D) = \ell$ and let $X_{n,\ell}$ be the set $\{\overline{d} \in \mathbb{Z}^n,\ o \leq d_1 \leq d_2 \leq \ldots \leq d_n < \ell\}$. On $X_{n,\ell}$ we introduce the following equivalence relation : we put $\overline{d} \sim \overline{d}'$ if and only if there exist $t, q_1, \ldots, q_n \in \mathbb{Z}$ and $\sigma \in S_n$ such that $d_i + t = \ell q_i + d'_{\sigma(i)}$, $i = 1, \ldots, n$. Let $C_n(D)$ be the quotient set $X_{n,\ell}/\sim$ and it is clear that this is a finite set (as ℓ is finite), so let $c_n(D)$ be the cardinality of $C_n(D)$.

9.1.3. Proposition. Let D be a graded division ring with $\ell = d(D) > o$ and let n be fixed natural number. The number of isomorphism classes of graded simple rings of size n over D is exactly in $c_n(D)$.

Proof : From Lemma 9.1.1. it follows that $\overline{d}\sim\overline{d'}$ yields $M_n(D)(\overline{d})\cong M_n(D)(\overline{d'})$. Conversely suppose that $M_n(D)(\overline{d})\cong M_n(D)(\overline{d'})$. Consider the graded D-modules : $V=D(-d_1)\oplus\ldots$ $\ldots\oplus D(-d_n)$, $W=K(-d_1')\oplus\ldots\oplus K(-d_n')$. Then $END_D(V)=END_D(W)$. Since D is principal we may use a graded version of Theorem 1.6 p.37 in [19] , to obtain the existence of : an isomorphism of degree o, $\alpha : D\to D$, an integer p and the α-isomorphism of degree o : $f : V(-p)\to W$. Therefore we get an α-isomorphism

$$D(-d_1-p)\oplus\ldots\oplus D(-d_n-p)\cong D(-d_1')\oplus\ldots\oplus D(-d_n').$$

Choose now $d''\in\mathbb{Z}^n$ such that $\overline{d+p}=\overline{\ell q}+\overline{d''}$ and with $d_i''<\ell$ for all $i=1,\ldots,n$. Then for any integer m there is a D-isomorphism of degree o; $D\cong D(-\ell m)$, hence, due to the existence of f, an α-isomorphism of degree o :

$$D(-d_1'')\oplus\ldots\oplus D(-d_n'')\cong D(-d_1')\oplus\ldots\oplus D(-d_n').$$

Since $D_1=D_2=\ldots=D_{\ell-1}=o$ there exists $\sigma\in S_n$ such that $d_i''=d_{\sigma(i)}'$ $1\leqslant i\leqslant n$. □

9.1.4. Example. Let D be a graded division ring. If $d(D)=1$ then $c_n(D)=1$. If $d(D)=2$ then $c_n(D)=[\frac{n}{2}]+1$.

9.1.5. Example. Let D be a trivially graded division ring and take $n\geqslant 2$. Let $m>o$ be an integer and consider the n-tupel $(o,o,\ldots,o,m)=\overline{m}$. Let $\dot{R}_\lambda=o$ for all $\lambda\neq o$, m and put :

$$R_o=\begin{bmatrix} D & \cdots & D & 0 \\ D & \cdots & D & 0 \\ \vdots & & \vdots & \vdots \\ D & & D & 0 \\ 0 & 0 & 0 & D \end{bmatrix} \qquad R_m=\begin{bmatrix} 0 & \cdots & & 0 \\ \vdots & & & \\ 0 & \cdots & & 0 \\ D & D & D & 0 \end{bmatrix}$$

Notice that there exists an infinity of classes of simple graded rings of size n over D.

9.2. Rings of Fractions and Goldie's Theorems.

Recall that if R is any ring, S a multiplicatively closed set in R, then the left ring of fractions $S^{-1}R$ with respect to S exists if and only if R and S satisfy:

i) If $s \in S$, $r \in R$ are such that $rs = o$ then there exists an $s' \in S$ such that $s'r = o$.

ii) For $r \in R$, $s \in R$ there exists $r' \in R$, $s' \in S$ such that $s'r = r's$.

These conditions are known as Ore's conditions on the left; usually, e.g. if R is left Noetherian one only has to check ii).

9.2.1. Lemma. Let R be a graded ring, S a multiplicatively closed subset of R consisting of homogeneous elements. Then S satisfies Ore's conditions on the left if and only if :

i) If $rs = o$ with $r \in h(R)$, $s \in S$ then there is an $s' \in S$ such that $s'r = o$.

ii) For any $r \in h(R)$, $s \in S$ there exist $r' \in h(R)$, $s' \in S$ such that $s'r = r's$.

Proof : Obviously the Ore conditions for S imply i) and ii). Conversely, consider $s \in S$, $r \in R$, write $r = r_1 + \dots + r_n$ with $\deg r_1 < \dots < \deg r_n$, $r_i \neq o$ $i = 1, \dots, n$. If $n = 1$ then the proof is finished. Now we proceed by induction on n, so there exist $r^1 \in R$, $s^1 \in S$ such that : $s^1(r_1 + \dots + r_{n-1}) = r^1 s$. Pick $s^2 \in S$, $r^2 \in R$ such that $s^2 r_n = r^2 s$. Consider $u \in S$, $v \in h(R)$ with $us^1 = vs^2 = t$ and put $ur^1 + vr^2 = w$, then $tr = ws$. For the second statement suppose $as = o$ with $a = a_1 + \dots + a_n$, $s \in S$. Thus $(a_1 + \dots + a_{n-1})s =$ and $a_n s = o$. The induction hypothesis yields the existence of $t_1 \in S$ such that $t_1(a_1 + \dots + a_{n-1}) = o$. However, since $t_1 a_n s = o$ also, there exists $t_2 \in S$ such that $t_2 t_1 a_n = o$, hence, putting $s' = t_1 t_2$ we get $s'a = o$. □

Let R be a graded ring and S a multiplicatively closed subset of h(R) satisfying Ore's conditions. Define a gradation on $S^{-1}R$ by putting :

$$(S^{-1}R)_\lambda = \{a/s, a \in h(R),\ s \in S;\ \deg a - \deg s = \lambda\} .$$

It is easy to see that $S^{-1}R$ thus becomes a graded ring. In a similar way, for any $M \in R\text{-gr}$, the graded module of fractions $S^{-1}M$ is obtained by equipping $S^{-1}M$ with the grading : $(S^{-1}M)_\lambda = \{m/s,\ m \in h(M), s \in S,\ \deg m - \deg s = \lambda\}$. Results about these constructions follow also from the general theory of graded localizations in II.

A graded ring which is of finite Goldie dimension in R-gr and which satisfies the ascending chain condition on graded left annihilators is called a graded Goldie Ring. If R is trivially graded then we get the usual notion of Goldie ring, cf. [11]. Let us mention Goldie's Theorem : a ring R has a semisimple (simple) Artinian classical ring of fractions if and only if R is a semiprime (prime) Goldie ring.
Exactly as in the ungraded case one can prove that, if R has a simple (semisimple) Artinian graded ring of fractions then R is a prime (semiprime) Goldie ring. However the converse is not true, as the following shows :

9.2.2. Example. Let k be a commutative field, R a graded ring such that $R^+ = k[X]$, $R^- = k[Y]$ and $XY = YX = 0$, i.e. $R_n = kX^n$, $R_0 = k$, $R_{-n} = kY^n$ for $n > 0$.
It may be easily verified that R satisfies :
i) Every $x \in h(R)$ with deg $x \neq 0$ is non-regular.
ii) The ideal $(X) + (Y)$ is essential in R and it does not contain any regular homogeneous element.
The ring R is a semiprime graded Goldie ring but it has no semisimple graded ring of fractions. Still we may prove the following :

9.2.3. Proposition. Let R be a semiprime graded Goldie ring which satisfies one of the following properties :
a. R has a central regular homogeneous element s with deg $s > 0$.
b. R is positively graded and minimal prime ideals of R do not contain R_+.
c. R is positively graded and R has a regular homogeneous element of positive degree.
d. Homogeneous elements of R of positive degree are nilpotent.
Then R admits a semisimple Artinian ring of fractions.

Proof : Let L be a nonzero graded left ideal of R. Under the hypothesis a,b,c, it is easy to establish that L contains a homogeneous element of positive degree which is not nilpotent. Under the hypothesis d, it follows that L contains a non-nilpotent element of degree o.
The essential part of the proof is to show that a left graded essential ideal contains

a regular homogeneous element. Under the assumptions a,b,c, we find that there is an $a_1 \in h(L)$ with deg $a_1 > o$ and $a_1^{\nu} \neq o$ $\nu = 1,\dots,\dots$. For some $n \in \mathbb{N}$: $\mathrm{Ann}(a_1^n) = \mathrm{Ann}(a_1^{n+1}) = \dots$. Put $b_1 = a_1^n$, then deg $b_1 > o$ and $\mathrm{Ann}(b_1) = \mathrm{Ann}(b_1^{\nu})$. If $\mathrm{Ann}(b_1) \neq o$ then there is an homogeneous $b_2 \in L \cap \mathrm{Ann}(b_1)$ with deg $b_2 > o$ and $\mathrm{Ann}(b_2) = \mathrm{Ann}(b_2^{\nu}), \nu = 1,2,\dots$. Proceeding in this way we obtain a direct sum $Rb_1 \oplus Rb_2 \oplus \dots$ of nonzero graded left ideals and since the Goldie dimension of R is finite, there exists an $r \in \mathbb{N}$ such that :

$$\mathrm{Ann}(b_1) \cap \dots \cap \mathrm{Ann}(b_r) = o \ .$$

Let $d_i = \deg b_i > o$, $1 \leq i \leq r$ and $d = d_1 \dots d_r$. Then $c = b_1^{d/d_1} + \dots + b_r^{d/d_r}$ is a homogeneous element of L with deg $c = d$. However $\mathrm{Ann}\ c = \bigcap_{i=1}^{r} \mathrm{Ann}(b_i^{d/d_i}) = \bigcap_{i=1}^{r} \mathrm{Ann}(b_i) = o$; so Rc is an essential graded left ideal, c is a regular element and deg $c > o$. Under the assumption d one may easily prove that L contains a regular homogeneous element of degree o. From here on the proof will be similar to the proof of Goldie's theorem's in the ungraded case. □

<u>9.2.4. Corollary</u>. Let R be a left and right Noetherian graded ring which is either positively graded or commutative. Let P be any prime ideal of R then either : $ht(P) = ht((P)_g)$ or $ht(P) = ht((P)_g) + 1$ (where ht(-) denotes the height of the ideal mentioned).

<u>Proof</u> : Suppose that Q is a prime ideal of R such that $(P)_g \subset Q \subset P$. Up to passing to $R/(P)_g$ we may assume that $(P)_g = o$ and hence that R is a prime ring.
Let S be the set of regular homogeneous elements of R. Clearly, R being Noetherian prime and graded, R is a graded Goldie ring hence, the foregoing proposition implies that $S^{-1}R$ exists and that it is a simple Artinian graded ring. From our results on the Krull dimension we retain : $K.\dim S^{-1} R \leq 1$. On the other hand $S^{-1}Q$ and $S^{-1}P$ are nonzero prime ideals of $S^{-1}R$, therefore $S^{-1}Q = S^{-1}P$. This means, becauxe of the right Noetherian property that $sP \subset Q$ for some $s \in S$, $s \neq o$ hence $P = Q$ or $s \in Q$; would imply $s \in (P)_g = o$ the latter is impossible hence $P = Q$ is the only remaining possibility. Now if $P = (P)_g$ then $ht(P)_g) = ht(P)_g)$ so suppose that $P \neq (P)_g$ and assume

that $ht(P) = n < \infty$. For $n = 1$, $(P)_g$ is a minimal prime ideal, i.e. $ht(P)_g = 0$. We proceed by induction on n. Let $P_0 \subset P_1 \subset \ldots \subset P_n = P$ be an ascending chain of length n of prime ideals contained in P. By the induction hypothesis we have $ht(P_{n-1}) = n-1$ if $P_{n-1} = (P_{n-1})_g$ but in this case $(P)_g = P_{n-1}$ and also $ht(P) = ht((P)_g) + 1$. On the other hand if $P_{n-1} \neq (P_{n-1})_g$ then $ht(P_{n-1}) = ht((P_{n-1})_g)+1$. Now $(P)_g \neq (P_{n-1})_g$ since otherwise P_{n-1} would be properly contained in P and $P_{n-1} \supset (P)_g$, so $ht((P)_g) > ht((P_{n-1})_g)$. Thus $ht((P)_g) \geq n-1$ and $ht(P) \leq ht((P)_g)+1$. □

9.2.5. Corollary. Let R be a Noetherian commutative graded ring. If P is a graded prime ideal with $ht(P) = n$, then there exists a chain of graded prime ideals : $P_0 \subsetneq P_1 \subsetneq \ldots \subset P_n = P$.

Proof : The statement is true for $n = 1$. If $ht(P) = n > 1$ then there is a chain of distinct prime ideals : $Q_0 \subsetneq Q_1 \subsetneq \ldots \subset Q_n = P$.
If Q_{n-1} is graded then application of the induction hypothesis finishes the proof. Suppose that Q_{n-1} is not graded and then the foregoing corollary yields that $ht((Q_{n-1})_g) = n-2$. The induction hypothesis asserts that we may find a chain of distinct graded prime ideals $P_0 \subset P_1 \subset \ldots \subset P_{n-2} = (Q_{n-1})_g$. Consider the graded ring $R/(Q_{n-1})_g = \overline{R}$ and denote by $\overline{P}, \overline{Q}_{n-1}$ the image of P, Q_{n-1} in $\overline{R}$. Choose a nonzero homogeneous element $\overline{a}$ of $\overline{P}$, hence $\overline{a} \notin \overline{Q}_{n-1}$. Let $\overline{P}_{n-1}$ be a minimal prime ideal containing $\overline{R}\,\overline{a}$. Then $\overline{P}_{n-1}$ is graded and $\overline{P}_{n-1} \neq \overline{P}$ because otherwise $ht(\overline{P}) \leq 1$ follows from the principal ideal theorem. Putting $\overline{P}_{n-1} = P_{n-1}/(Q_{n-1})_g$ we find a chain of distinct prime ideals : $P_0 \subset P_1 \subset \ldots \subset P_{n-1} \subset P_n = P$. □

9.2.6. Corollary. Let R be a Noetherian commutative graded ring and let I be an ideal of R, then :

$$ht(I) \leq ht(I^{\sim}) \ , \ ht(I) \leq ht(I_{\sim}).$$

Proof : Since $ht(I) = ht(rad(I))$ and $(rad(I))^{\sim} \subset rad(I^{\sim})$ (see Lemma 1.1.) we may reduce the proof to the case where I is a semiprime ideal, $I = P_1 \cap \ldots \cap P_s$, P_i prime ideals of R. Then $P_1 \ldots P_s \subset I$ yields $\tilde{P_1}.\tilde{P_2} \ldots \tilde{P_s} \subset I^{\sim}$ (Lemma 1.1.). Let Q be a prime ideal containing $I^{\sim}$ such that $ht(I^{\sim}) = ht(Q)$. For some i, $P_i^{\sim} \subset Q$, but $I \subset P_i$ hence we may assume that $I = P$ is a prime ideal in proving the statement. If P is graded then $P = P^{\sim}$. If P is not graded then $ht(P) = ht((P)_g) + 1$. Clearly, $(P)_g \subset P^{\sim}$ and $(P)_g \neq P^{\sim}$. It results from this that $ht((P)_g) < ht(P^{\sim})$ and therefore $ht(P) \leq ht(P^{\sim})$. In a similar way, $ht(I) \leq ht(I_{\sim})$ may be established. □

II.10. PRIMARY DECOMPOSITION

Let R be a graded ring, P a graded ideal of R. It is obvious that P is a prime ideal of R if and only if for $a,b \in h(R)-P$, $aRb \not\subset P$. We write spec R, $\text{spec}_g R$ for the set of prime ideals of R, resp. graded prime ideals of R. In view of Lemma 1.5., the minimal prime ideals of R are graded and the radical of a graded ideal is also a graded ideal. If $\underline{M}$ is a nonzero R-module the a prime ideal P of R is said to be associated to $\underline{M}$ if there exists a nonzero submodule M' of M such that $P = \text{Ann } M' = \text{Ann } \underline{M}''$ for every nonzero submodule $\underline{M}''$ of $\underline{M}'$.

Let Ass $\underline{M}$ be the set of prime ideals of R which are associated to $\underline{M}$. In case R is a commutative ring then $P \in \text{spec } R$ is associated to $\underline{M}$ if and only if there is an $x \in \underline{M}$, $x \neq o$, such that $P = \text{Ann } x$. A well known result in ring theory states that, if R is a left Noetherian ring and $\underline{M} \neq o \in R\text{-mod}$, then Ass $\underline{M} \neq o$.

10.1. Theorem. Let R be a graded ring, M nonzero in R-gr.
Suppose that $P \in \text{Ass } \underline{M}$, then :

1. P is graded.
2. There is an $x \in h(M)$, $x \neq o$, such that $P = \text{Ann } Rx$.
3. If R is a fully left bounded left Noetherian ring (cf. [36] for definition and properties of these rings) then there exists $x \in h(M)$, $x \neq o$, such that $P = \text{Ann } Rx = \text{Ann } \underline{M}'$ for any nonzero submodule $\underline{M}' \subset Rx$.

Proof : 1. By hypothesis $P = \text{Ann } \underline{M}'$ for some nonzero R-submodule $\underline{M}'$ of $\underline{M}$. Take $\lambda \in P$, $x \in \underline{M}'$ and write $\lambda = \sum'_{i \in \mathbb{Z}} \lambda_i$, $x = \sum'_{j \in \mathbb{Z}} x_j$. Let m, resp. n, be the largest integer such that $\lambda_m \neq o$, resp. $x_n \neq o$. Pick $a \in h(R)$. Since $\lambda a \in P$ we have $ax = o$, hence $\lambda_m a x_n = o$, implying $\lambda_m R x_n = o$.

Let $b\in h(R)$. From $\lambda bx=o$ we deduce that. $\lambda_m bx_{n-1}+\lambda_{m-1}bx_n=o$. Replacing b by $b\lambda_m a$ in the above equality yields $\lambda_m b\lambda_m ax_{n-1}=o$, hence $(\lambda_m R)^2 x_{n-1}=o$. The number of nonzero components of x being finite, it follows that $(\lambda_m R)^p\, x_i=o$ for all $x_i\neq o$ appearing in x, for some $p\in\mathbb{Z}$. Consequently $(\lambda_m R)^p Rx=o$, thus $(\lambda_m R)^p\subset P$ and $\lambda_m R\subset P$. Thus $\lambda-\lambda_m\in P$ and repetition of this procedure yields that $\lambda_i\in P$ for all λ_i appearing in the decomposition of λ i.e. P is graded.

2. Take $x\neq o\in \underline{M}'$. Then $P=\text{Ann } Rx$. Set $x=x_1+\ldots+x_k$ where $x_j\neq o$ for all j, $\deg x_1<\ldots<\deg x_k$.

Let J_i be $\text{Ann } Rx_i$, $1\leq i\leq k$. It is easy to verify that $P=J_1\cap\ldots\cap J_k$, thus $P=J_i$ for some i, therefore $P=\text{Ann } Rx_i$.

3. According to 2 we can find $x\in h(M)$, $x\neq o$, such that $P=\text{Ann } Rx$. Put $N=Rx$. By our assumption $\underline{N}$ is a faithful R/P-module, where R/P is a fully left bounded left Noetherian ring. By [19], Lemma 2.1 and Proposition 1.4., it follows that $\underline{N}$ is a nonsingular R/P-module. However N is graded hence there exists $y\in h(N)$, $y\neq o$, such that $\text{Ann}_{R/P}\, y$ is not an essential left ideal of R/P, meaning that there may be found a graded left ideal isomorphic to a graded submodule A of N. Let $z\in h(A)$, $z\neq o$. Clearly, $P=\text{Ann}_R Rz=\text{Ann}_R \underline{M}'$ for any nonzero submodule $\underline{M}'$ of Rx. □

Let $\underline{M}\in R\text{-mod}$; let $\underline{M[X]}$ be the $R[X]$-module of polynomials with coefficients in $\underline{M}$. We have $M[X]=R[X]\otimes_R \underline{M}$. We know that : $M[X]_i=\{mX^i, m\in M\}$, defines a grading on $M[\underline{X}]$.

<u>10.2. Theorem.</u> Let R be a graded ring, $M\in R\text{-mod}$. Then :

$$\text{Ass}_{R[X]} M[X]=\{P[X],\ P\in \text{Ass } \underline{M}\}.$$

<u>Proof</u> : If $P\in\text{Ass } \underline{M}$ then $P=\text{Ann}_R \underline{M}'$ for some $\underline{M}'\subset\underline{M}$. Then it is clear that $P[X]=\text{Ann}_{R[X]} M'[X]$. Let $m'X^\nu$ be a nonzero element of $h(M'[X])$. Since $P=\text{Ann}_R Rm'$ it follows that $P[X]=\text{Ann}_{R[X]} R[X].m'X^\nu$. Hence $P[X]\in\text{Ass}_{R[X]}\underline{M[X]}$. Conversely, let $Q\in\text{Ass}_{R[X]}\underline{M[X]}$. Theorem 10.1.,2., implies that there exists a nonzero homogeneous element mX^ν with $Q=\text{Ann}_{R[X]} R[X].mX^\nu$. Put $P=\text{Ann}_R Rm$. It is obvious that $Q=P[X]$, P is prime and $P\in\text{Ass}_R M$. □

Let $\underline{N} \subset \underline{M}$ be R-modules. We say that N is a primary submodule of M if Ass $\underline{M/N}$ is a singleton, and N is said to be a P-primary submodule of M if Ass $\underline{M/N} = \{P\}$. If R is Noetherian commutative and $\underline{M}$ is of finite type then $\underline{N}$ is a primary submodule of $\underline{M}$ if and only if N is a classical primary submodule, (cf.[3],ch.4).

10.3. Lemma. Let R be a left Noetherian ring, $M \in R$-mod and $P \in$ spec R. The following statements are equivalent :

1. O is a P-primary submodule of M.
2. The set $(0 : P)_M = \{x \in M,\ Px = 0\}$ is an essential submodule of M and P contains any ideal which annihilates a nonzero submodule of M.

Proof : $1 \Rightarrow 2$. If X is a nonzero submodule of M then Ass $X = \{P\}$ implies that $X \cap (0:P)_M \neq 0$. If I is an ideal which annihilates a nonzero submodule $M' \subset M$. Since, Ass $M' = \{P\}$, there is an $M'' \subset M'$ with $M'' \neq 0$ and $P = \text{Ann } M''$. However $IM' = 0$ yields $IM'' = 0$, hence $I \subset P$.

$2 \Rightarrow 1$. Since R is left Noetherian, Ass $M \neq Q$. Let $Q \in$ Ass M. From our assumptions it follows that $Q \subset P$. Consider $M' \neq 0$, $M' \subset M$, with $Q = \text{Ann } M'$. The fact that $M' \cap (0:P)_M \neq 0$ entails : $P \subset \text{Ann}(M' \cap (0:P)_M) = Q$. Therefore $P = Q$. □

10.4. Remark. If R is a graded ring, $M \in R$-gr then in Lemma 10.3 we may replace "ideal" in statement 2 by graded ideal.

10.5. Lemma. Let R be a graded ring and $M \in R$-gr. Let $\underline{N} \subset \underline{X}$ be submodules of $\underline{M}$. If $\underline{X}/\underline{N}$ is essential in $M/(N)_g$ then $X^{\sim}/(N)_g$ (resp. $X_{\sim}/(N)_g$) is essential in $M/(N)_g$. In particular, if $\underline{X}$ is a left essential submodule of $\underline{M}$ then $X^{\sim}$ and $X_{\sim}$ are left esstenial in

Proof : Take $x \in h(M) - (N)_g$ and let $\bar{x}$ be its image in $M/(N)_g$. Since $x \notin N$, its image $\hat{x}$ is M/N is nonzero. By the assumption we find $\lambda \in R$ such that $\lambda\hat{x} \in \underline{X}/\underline{N}$ and $\lambda\hat{x} \neq 0$. Writing $\lambda = \lambda_1 + \dots + \lambda_n$ with deg $\lambda_1 < \dots <$ deg λ_n we find that $\lambda x = \lambda_1 x + \dots + \lambda_n x$ and we may assume that $\lambda_n x \notin \underline{N}$ because otherwise we may replace λ by $\mu = \lambda - \lambda_n$ while $\lambda\hat{x} = \lambda\hat{x}$. Since $\lambda_n x \notin \underline{N}$ we have $\lambda_n x \neq 0$. On the other hand, $\lambda\hat{x} \in \underline{X}/\underline{N}$ yields $\lambda x \in \underline{X}$, hence $\lambda_n x \in X^{\sim}$. But $\lambda_n x \notin (N)_g$, therefore $\lambda_n \bar{x} \in X^{\sim}/(N)_g$ and $\lambda_n \bar{x} \neq 0$. Lemma I.3.3.13. finishes the proof. The statement about $X_{\sim}/(N)_g$ follows in a similar way. □

10.6. Theorem. Let R be a left Noetherian graded ring, $M \in R\text{-gr}$ and N a P-primary submodule of M. Then $(N)_g$ is $(P)_g$-primary submodule of M.

Proof : Set $X/N = (0:P)_{M/N}$; clearly this is left essential in M/N. By the foregoing lemma we deduce that $X^{\sim}/(N)_g$ is essential in $M/(N)_g$. $(P)_g X^{\sim} \subset (N)_g$ follows from $PX \subset N$, hence $X^{\sim}/(N)_g \subset (0:(P)_g)_{M/(N)_g}$ and therefore the latter left module is essential in $M/(N)_g$. Consider an ideal I which is graded and such that it annihilates a nonzero $Y/(N)_g$, i.e. $IY \subset (N)_g$. Since I is graded : $IY^{\sim} \subset (N)_g$ and since $Y \neq (N)_g$ we have $(N)_g \neq Y^{\sim}$ i.e. $Y^{\sim}/(N)_g \neq 0$. Therefore, we may assume that I annihilates the graded nonzero submodule $Y/(N)_g$ of $M/(N)_g$. As before we may deduce that $IY \subset (N)_g$ i.e. $IY \subset N$. The fact that Y is graded yields $Y \not\subset N$ and we arrive at the conclusion that $I(Y+N/N) = 0$ with $Y+N/N \neq 0$. By hypothesis $I \subset P$ and $I \subset (P)_g$ follows. Then apply Lemma 10.3. □

10.7. Corollary. Let R be a left Noetherian ring and M a nonzero left R-module. The o is a P-primary submodule of M if and only if o is P[X]-primary in M[X].

Proof : Follows from 10.6 and 10.2. □

A finite family $(N_i)_{i \in I}$ of primary submodules of M is a primary decomposition for the submodule N of M if $N = \bigcap_{i \in I} N_i$. The decomposition is said to be a reduced decomposition if the following requirements are fulfilled :

i) $\bigcap_{j \neq i} N_j \not\subset N_i$ for all $i \in I$.

ii) If Ass $M/N_i = P_i$ then the ideals P_i are mutually disjoint.

For more detail about primary decomposition cf. [3], chapter 4. It is a well known result that, if R is left Noetherian and if $M \in R\text{-mod}$ is of finite type, then any submodule of M admits a reduced primary decomposition.

10.8. Corollary. Let R be a left Noetherian graded ring, M a graded R-module of finite type, N a graded submodule of M such that $N = \bigcap_{i \in I} N_i$ is a primary decomposition of N in M. Then we also have the following :

1. $N = \bigcap_{i\in I} (N_i)_g$ is a primary decomposition of N in M.

2. If the primary decomposition $N = \bigcap_{i\in I} N_i$ is reduced then the decomposition

$N = \bigcap_{i\in I} (N_i)_g$ is reduced and for any $i \in I$, Ass M/N_i = Ass $M/(N_i)g$.

Proof : 1. Follows from Theorem 10.6.

2. By Theorem 10.6, Ass M/N_i = Ass $M/(N_i)_g$ for all $i \in I$.

Suppose that $\bigcap_{j\neq i} (N_j)_g \subset (N_i)_g$, then $N = \bigcap_{j\neq i} (N_j)_g$. There exists an injective homomorphism, $\varphi : M/N \to \bigoplus_{j\neq i} M/(N_j)_g$, thus we may conclude that :

$$\text{Ass } M/N \subset \text{Ass } \bigoplus_{j\neq i} M/(N_j)_g = \bigcup_{j\neq i} \text{Ass } M/(N_j)_g = \bigcup_{j\neq i} \text{Ass } M/N_j ,$$

contradiction.

Hence the decomposition $N = \bigcap_{i\in I} (N_i)_g$ is reduced. □

By a classical P-primary submodule of $M \in R$-mod we mean a submodule N of M which is P-primary and has the property that $P^n M \subset N$ for some natural number n Theorem 10.6 implies that if $M \in R$-gr, and N is a classical P-primary submodule of M, then $(N)_g$ is a classical $(P)_g$-primary submodule of M. Corollary 10.8 still holds if we replace "primary" by "classical primary". That, for a left Noetherian ring R and an R-module M of finite type it is not true in general that 0 has a classical primary decomposition in M, is just old hat.

Finally let us point out that all results in this section remain valid if R is a graded ring of type Δ where Δ is any commutative torsion free group.

II.11 EXTERNAL HOMOGENIZATION.

Let R be a graded ring. The ring of polynomials $R[T]$ may be turned into a graded ring by putting : deg T = 1, $R[T]_n = \{ \sum_{i+j=n} r_i T^j, r_i \in R_i\}$. In the same way we build the module of polynomials $M[T]$ starting from an $M \in R$-gr. It is easily checked that $M[T]$ is an $R[T]$-module and that $M[T] \cong R[T] \otimes_R M$, where the tensor product is graded in the usual way. If we decompose $x \in M$ into homogeneous components, $x = x_{-m} + \dots + x_0 + \dots + x_n$, then we may associate to it a homogeneous element x^* of $M[T]$ which is given by :

$$x^* = x_{-m}T^{m+n} + x_{1-m}T^{m+n-1} + \dots + x_0T^n + \dots + x_n .$$

We say that x^* is the homogenized of x. If a is a homogeneous element of $M[T]$, say $u = u_{-k}\,T^{k+j} + \dots + u_0T^j + \dots + u_j$, with $u_i \in M, -k \leqslant i \leqslant j$, then $u_* = u_{-k} + \dots + u_0 + \dots + u_j \in M$ is said to be the dehomogenized of u. We have the following :

i) If $f \in R[T]$, $u \in M[T]$ are both homogeneous then $(fu)_* = f_*u_*$.

ii) If $u,v \in h(M[T])$ have the same degree then $(u+v)_* = u_* + v_*$.

iii) If $x \in M$, $(x^*)_* = x$.

iv) If $u \in M[T]$ is homogeneous, then $(u_*)^*T^k = u$, where $k = \deg u - \deg(u_*)^*$.

If $\rho \in R$, $m \in M$ then $d(\rho)$ or $d(m)$ will denote the highest degree appearing in a homogeneous decomposition of ρ or m.

v) $\rho^*m^* = T^k(\rho m)^*$, where $k = d(\rho) + d(m) - d(\rho m)$.

vi) Let $x,y \in M$ and suppose $d(x) > d(y)$, then : $T^k(x+y)^* = x^* + T^{\ell}y^*$, where $\ell = d(x) - d(y)$ and $k = \max\{d(x), d(y)\} - d(x+y)$.

If N is an R-submodule of M then by N^* we mean the submodule of $M[T]$ generated by the n^*, $n \in N$. Of course N^* is a graded submodule of $M[T]$, it is called the homogenized of N. Any $n \in N^*$ is of the form $T^{\nu}n^*$, $n \in N$, $\nu \geqslant 0$.

To a graded $R[T]$-submodule L of $M[T]$ we may associate $L_* = \{u_*, u \in h(L)\}$ and from the properties mentioned above it follows that L_* is an R-submodule of M. The correspondence $N \to N^*$ satisfies :

1) $(N^*)_* = N$.

2) If $L \subsetneq N$ then $L^* \subsetneq N^*$.

3) If N is a graded submodule of M then $N^* = N[T]$.

4) If I is a left ideal of R then $(IN)^* = I^*N^*$.

5) If $x \in M$ then $(N : x)^* = N^* : x^*$

6) $(\bigcap_{i\in I} N_i)^\star = \bigcap_{i\in I} N_i^\star$.

The correspondence $L \to L_\star$ satisfies :

1) $(L_\star)^\star \supset L$.

2) If $L \subset L'$ then $L_\star \subset L'_\star$.

3) $(\bigcap_{\in J} L_i)_\star = \bigcap_{i\in J} (L_i)_\star$.

4) If L is a graded left ideal of $R[T]$ then $(JL)_\star = J_\star L_\star$.

5) If $u \in M[T]$ is homogeneous then $[L:u]_\star = L_\star : u_\star$.

6) $(\sum_{i\in J} L_i)_\star = \sum_{i\in J} (L_i)_\star$.

From this it is clear that we may define a functor,

$$E : R[T]\text{-gr} \to R\text{-mod} ,$$

such that $E(M) = R[T]/(T-1) \otimes_{R[T]} M \cong M/(T-1)M$.

11.1. Lemma. With notations as above we have :

1. E is an exact functor.
2. If N is an R-module (of finite type, of finite presentation then there exists an $M \in R[T]$-gr (of finite type, of finite presentation) such that $E(M) \cong N$.
3. If R has negative grading and if P is a projective R-module, then there is a projective object Q in $R[T]$-gr such that $E(Q) \cong P$.

Proof : 1. Right exactness of E is obvious. Take $M \in R[T]$-gr and let M be a graded submodule of M. For left exactness of E it will be sufficient to have that $M' \cap (T-1)M = (T-1)M'$. Pick $x \in M' \cap (T-1)M$, then $x = (T-1)m \in M'$ with $m \in M$. $m = m_{-t} + \dots + m_0 + \dots + m_s$ with $m_i, -t \leqslant i \leqslant s$, being the component of degree i of m. From $(T-1)m \in M'$ we get :

$$-m_{-t} + Tm_{-t} - m_{-t+1} + Tm_{-t+1} - \dots - m_s + Tm_s \in M' .$$

Because M' is graded, $m_{-t} \in M'$; then $m_{-t+1} \in M'$ and so on till we find that $m \in M'$ i.e. $x \in (T-1)M'$.

2. Write $N = F/L$ where $F = R^{(I)}$ is a free R-module equipped with the grading induced by the grading of R. The $R[T]$-module $F[T]$ is free graded. If $L^{\star}$ is the homogenized of L in $F[T]$ then we put $M = F[T]/L^{\star}$. One easily verifies that $M/(T-1)M \cong N$.

3. Let $P = F/L$ be projective i.e. $F = L \oplus K$ with $P \cong K$. Since R is negatively graded $F_i = 0$ for all $i > 0$. Take the canonical basis $(f_i)_{i \in J}$ of F, i.e. $f_i = (0, \ldots, 1, 0, \ldots, 0)$, then $(f_i)_{i \in J}$ also is an $R[T]$-basis for $F[T]$. Since $F = L \oplus K$ we may write $u_i = x_i + y_i$ with $x_i \in L$, $y_i \in K$. However $\deg u_i = 0$ and $F_i = 0$ for $i > 0$ yields that either $d(x_i + y_i) = d(x_i) = 0$ or $d(x_i + y_i) = d(y_i) = 0$. In both cases $(x_i + y_i)^{\star} = x_i^{\star} + y_i^{\star}$ and hence $u_i = x_i^{\star} + y_i^{\star}$. Thus $L^{\star} + K^{\star} = F^{\star} = F[T]$. Since $L^{\star} \cap K^{\star} = 0$ it follows that $F[T] = L^{\star} \oplus K^{\star}$ and therefore, if $Q \cong F[T]/L^{\star}$ then Q is projective in $R[T]$-gr while $E(Q) \cong F/(L^{\star})_{\star} = F/L = K \cong P$. □

Note. If 3. holds for positively graded rings then it would be possible to derive from it an easy proof for Serre's conjecture and it would also yield an affirmative answer to the Bass-Quillen conjecture. Again from this one can deduce that the positively graded analoge of 3. must be false for non-commutative R. For commutative R the question remains open; an affirmative answer would entail deep results in K-theory.

11.2. Theorem. Let R be a graded ring, then :

$$\text{gr.gl.dim } R \leq \text{gl.dim } R \leq 1 + \text{gr.gl.dim } R.$$

If R has a limited grading then :

$$\text{gr.gl.dim } R = \text{gl.dim } R .$$

Proof : 1. The inequality gr.gl.dim $R \leq$ gl.dim R is just Corollary 7.7. Now if $M \in R[T]$-gr then we have the following exact sequence in $R[T]$-mod :

$$0 \to M[T] \xrightarrow{i} M[T] \xrightarrow{j} M \to 0$$

where i and j are homogeneous, i is multiplication on the right by T-1. Consequently, gr.gl.dim $R \leq 1+$gr.gl.dim R. If $N \in R$-mod then let $M \in R[T]$-gr be such that $N \cong E(M)$. Exactness of E entails that $\text{gr.dim}_R N \leq \text{gr.p.dim}_{R[T]} M \leq \text{gr.gl.dim } R[T]$ hence gl.dim $R \leq$ gr.gl.dim R.

2. This part follows directly from Corollary 7.7. □

11.3. Theorem. Let R be a left Noetherian positively graded ring and assume that gl.dim $R<\infty$. Let $P\in R$-mod be projective and of finite type, then there exist projective R_0-modules of finite type Q and Q' such that :

$$P \oplus (R \underset{R_0}{\otimes} Q) \cong R \underset{R_0}{\otimes} Q' .$$

Proof : Choose $M \in R[T]$-gr such that $E(M) \cong P$, because of 11.1 M may be chosen to be of finite type. Since gl.dim $R<\infty$ we obtain the following exacts sequence in $R[T]$-gr.

$$(*) \quad 0 \to Q_n \to \dots \to Q_1 \to Q_0 \to M \to o$$

where $Q_0, Q_1 \dots Q_n$ are projective and of finite type in $R[T]$-gr. The grading of $R[T]$ is positive so Nakayama's lemma may be applied to deduce that any projective object of finite type in $R[T]$-gr is of the form $R[T] \underset{R_0}{\otimes} S$ where S is a projective R_0-module of finite type. Substituting $Q_i = R[T] \underset{R_0}{\otimes} S_i$ in (*) and then applying the functor E to the sequence, right exactness of E yields an exact sequence in R-mod :

$$o \to E(Q_n) \xrightarrow{d_n} E(Q_{n-1}) \xrightarrow{d_{n-1}} \dots \xrightarrow{d_2} E(Q_1) \xrightarrow{d_1} E(Q_0) \xrightarrow{d_0} P \to 0$$

where $E(Q_i) \cong Q_i/(T-1)Q_i \cong R \underset{R_0}{\otimes} S_i$.

By the projective of P we retain from all this that : $P \oplus \mathrm{Ker}\, d_0 \cong R \underset{R_0}{\otimes} S_0$;

$\mathrm{Ker}\, d_0 \oplus \mathrm{Ker}\, d_1 \cong R \underset{R_0}{\otimes} S_1 ; \dots ; R \underset{R_0}{\otimes} S_{n-1} \cong \mathrm{Ker}\, d_{n-1} \oplus R \underset{R_0}{\otimes} S_n$.

So we find that there exist projective R_0-modules Q and Q' such that $P \oplus (R \underset{R_0}{\otimes} Q) \cong R \underset{R_0}{\otimes} Q'$. □

11.4. Theorem. Let R be a left Noetherian ring with gl.dim.$R<\infty$. Let P be a projective left $R[T_1,\dots,T_n]$-module of finite type. There exist projective R-modules Q and Q', both of finite type, such that : $Q[T_1,\dots,T_n] \oplus P \cong Q'[T_1,\dots,T_n]$.

Proof : Apply Lemma 11.3 to $R[T_1,\dots,T_n]$. □

Note that in K-theoretic wording the above theorem states that $K_0(R) \cong K_0(R[T_1,\dots,T_n])$.

II.12 GRADED RINGS AND MODULES OF QUOTIENTS

12.1. Torsion Theories over Graded Rings.

Let C be a Grothendieck category (for general torsion theory C need only be abelian,). A torsion theory in C is a couple (T,F) of nonvoid classes of objects of C such that :

TT1 : $T \cap F = \{o\}$

TT2 : T is closed under homomorphic images in C

TT3 : F is closed under taking subobjects.

TT4 : For any $M \in C$ there is a subobject $t(M)$ of M such that $t(M) \in T$, $M/t(M) \in F$.

A torsion theory (T,F) is said to be hereditary if T is closed under taking subobjects as well. We will only consider hereditary torsion theories here. A kernel functor on C is a left exact subfunctor of the identity in C. A kernel functor κ is said to be idempotent if and only if $\kappa(M/\kappa(M)) = o$ for all $M \in C$. In this terminology, elementary results in torsion theory yield that there is a one-to-one correspondence between idempotent kernel functors κ in C and torsion theories in C. These also correspond in a one-to-one way to Gabriel-filters (or-topologies) in a chosen generator G for C; such a Gabriel-filter in G will then consist of the subobjects L of G such that $G/L \in T$, in casu : $\kappa(G/L) = G/L$ because the torsion class of a torsion theory associated to an idempotent kernel functor κ is given as the class of $M \in C$ such that $\kappa(M) = M$, while $M \in F$ if and only if $\kappa(M) = o$. For more detail about torsion theory cf [9], [10], [32].

Notation. In the sequel (T,F) will always be a herditary torsion theory in C and κ will be the corresponding idempotent kernel functor on C.

Evidently we wish to apply these techniques to R-gr = C. Since R-gr actually is a Grothendieck category this presents no real problems, however R is not a generator for R-gr so the beautiful theory as developped in case $G = R\text{-mod}$ does not generalize completely. Moreover we will encounter (solvable) problems in relating torsion theories in R-gr and torsion theories in R-mod.

12.1.1. Lemma. Let R be a graded ring, $(\mathcal{T},\mathcal{F})$ a torsion theory in R-gr. The following statements are equivalent.

1. If $M \in \mathcal{T}$ then $M(n) \in \mathcal{T}$ for any $n \in \mathbb{Z}$.
2. If $N \in \mathcal{F}$ then $N(m) \in \mathcal{F}$ for any $m \in \mathbb{Z}$.

Proof : Easy. □

12.1.2. Example. Let R be a positively graded ring. Consider for $\mathcal{T}$ the class of $M \in R\text{-gr}$ such that $M_i = o$ if $i < o$ and take $\mathcal{F}$ to be the class of $M \in R\text{-gr}$ such that $M_i = o$ if $i \geqslant o$. Clearly $(\mathcal{T},\mathcal{F})$ is a hereditary torsion theory which does not have the properties mentioned in Lemma 12.1.1. On the other hand it is not difficult to generate torsion theories which do have that property by adding to $\mathcal{T}$ all suspensions of objects from $\mathcal{T}$.

A torsion theory satisfying the equivalent conditions of Lemma 12.1.1. is said to be a rigid torsion theory. A kernel functor is rigid of the corresponding torsion theory is. For any torsion theory $(\mathcal{T},\mathcal{F})$ in R-gr we have that $\mathrm{Hom}_{R\text{-gr}}(M,N) = o$ for every $M \in \mathcal{T}$, $N \in \mathcal{F}$. However it should be noted that $\mathrm{HOM}_R(M,N) = o$ for all $M \in \mathcal{T}$, $N \in \mathcal{F}$, is equivalent to $(\mathcal{T},\mathcal{F})$ being rigid.

12.1.3. Lemma. A kernel functor κ in R-gr is rigid if and only if $\kappa(M(n)) = \kappa(M)(n)$ for all $n \in \mathbb{Z}$.

Proof : Suppose that κ is rigid, then $\kappa(M)(n) \subset \kappa(M(n))$. But $(M/\kappa(M))(n) = M(n)/\kappa(M)(n)$ is κ-torsion free i.e. in $\mathcal{F}$, therfore $\kappa(M(n))/\kappa(M)(n) = o$.
Conversely, if $M \in \mathcal{T}$ then $\kappa(M) = M$ i.e. $\kappa(M(n)) = M(n)$ follows and thus $M(n) \in \mathcal{T}$ for all $n \in \mathbb{Z}$. □

A non-empty set of graded left ideals $\mathcal{L}$ of R a graded filter if the following requirements are fulfilled :

F_1 : If $L \in \mathcal{L}$ and L_1 is a graded left ideal of R such that $L \subset L_1$ then $L_1 \in \mathcal{L}$.

F_2 : If $L_1, L_2 \in \mathcal{L}$ then $L_1 \cap L_2 \in \mathcal{L}$

F_3 : If $L \in \mathcal{L}$ then $[L : x] \in \mathcal{L}$ for all $x \in h(R)$.

F_4 : If $L_1 \in \mathcal{L}$ and $[L : x] \in \mathcal{L}$ for all $x \in h(L_1)$ then $L \in \mathcal{L}$.

As a consequence of the fact that R is not the generator of R-gr, we cannot expect that there is going to be a one-to-one correspondence between graded filters in R and hereditary torsion theories in R-gr. However we do have the following result :

12.1.4. Lemma. There is a bijective correspondence between, on-one hand the hereditary rigid torsion theories in R-gr, and on the other hand, rigid idempotent kernel functors on R-gr. The latter correspond further in a bijective way to the set of graded filters in R.

Proof : This is mere modification of the proof in the ungraded case. □

The sequel of this section is concerned with the characterization of those kernel functors on R-gr that may be obtained from kernel functors on R-mod. An idempotent kernel functor $\underline{\kappa}$ on R-mod is said to be graded if the filter $\mathcal{L}(\underline{\kappa})$ of $\underline{\kappa}$ possesses a cofinal set of graded left ideals.

12.1.5. Lemma. Let R be a graded ring and let $\underline{\kappa}$ be a graded kernel functor on R-mod. If M is a graded left R-module then $\underline{\kappa}(M)$ and $M/\underline{\kappa}(M)$ are graded left R-modules, the canonical R-morphism $M \to M/\underline{\kappa}(M)$ may be considered as a graded morphism of degree o. Furthermore : $\underline{\kappa}$ induces a kernel functor κ on R-gr which is rigid and idempotent, the torsion class $T(\kappa)$ consists of graded R-modules N such that $\underline{\kappa}(\underline{N}) = \underline{N}$, whereas $F(\kappa)$ consists of $N \in$ R-gr, $\underline{\kappa}(\underline{N}) = o$.

Proof : If we establish that $\underline{\kappa}(M)$ is graded then it will follow immediatly that

$\underline{M}/\underline{\kappa}(\underline{M})$ is graded and that $\underline{M}\to\underline{M}/\underline{\kappa}(\underline{M})$ is a graded morphism of degree o. Since $\underline{\kappa}$ is a graded kernel functor on R-mod, $x\in\underline{\kappa}(\underline{M})$ means that $Lx=o$ for some graded left ideal $L\in\mathcal{L}(\underline{\kappa})$. Write $x=\sum_i x_i$, $L=\sum_n L_n$. Then $L_n x=o$ for each $n\in\mathbb{Z}$, hence $L_n x_i=o$ for each $n,i\in\mathbb{Z}$. Thus $x_i\in\underline{\kappa}(M)$ for all $i\in\mathbb{Z}$ i.e. $\underline{\kappa}(M)$ is a graded submodule of M, which we will denote also by $\underline{\kappa}(M)$. Furthermore, if $T(\kappa)$ and $F(\kappa)$ are as stated then one easily verifies TT_1 till TT_4, so (T,F) is a hereditary torsion theory in R-gr which is easily seen to be rigid. The kernel functor κ corresponding to this torsion theory is a rigid kernel functor on R-gr such that $\kappa(M)=\underline{\kappa}(M)$ holds for every $M\in$ R-gr. □

As in any Grothendieck category we define the <u>torsion theory cogenerated in</u> R-gr by an object $M\in$ R-gr as follows : the torsion class $T(\kappa_M)$ consists of the $N\in$ R-gr. such that $\mathrm{Hom}_{R\text{-gr}}(N,E^g(M))=o$, where $E^g(M)$ is the injective hull of M in R-gr. If κ_M is a rigid kernel functor then $N\in T(\kappa_M)$ yields $\mathrm{HOM}_R(N,E^g(M))=o$ or $\mathrm{Hom}_{R\text{-gr}}(N,\bigoplus_n E^g(M)(n))=o$. Thus the rigid torsion theories amongst the κ_M, $M\in$ R-gr, may be thought of as being the torsion theories cogenerated by a suspension-invariant M. Let $\kappa_M(n)$ denote the torsion theory cogenerated in R-gr by M(n). The inf κ of torsion theories $\{\kappa_i, i\in I\}$ in R-gr is defined by taking for $T(\kappa)$ the intersection of the $T(\kappa_i)$, we denote this by $\kappa=\wedge\kappa_i$. It is then clear that $\wedge_n\kappa_M(n)$ is a rigid torsion theory which is cogenerated by $\bigoplus_n E^g(M)(n)=E^g(\bigoplus_n M(n))$. In particular, if we define $\kappa_1\leq\kappa_2$ on the lattice of hereditary torsion theories in R-gr by $T(\kappa_1)\subset T(\kappa_2)$ then a rigid torsion theory $\kappa\leq\kappa_M$ for some $M\in$ R-gr if and only if $\kappa\leq\wedge_n\kappa_M(n)$. The kernel functor $\wedge_n\kappa_M(n)$ is called the <u>rigid kernel functor associated or cogenerated</u> by $M\in$ R-gr., we will denote it by κ^r_M.

<u>12.1.6. Lemma.</u> Let $M\in$ R-gr and let $\kappa_{\underline{M}}$ denote the kernel functor in R-mod cogenerated by $\underline{M}\in$ R-mod then : $\mathcal{L}(\kappa_{\underline{M}})=\{J\subset\underline{R}$, for all $x\in M$, $x\neq o$, and $a\in R$, $(J:a)\not\subset\mathrm{Ann}_R x\}$
$\mathcal{L}(\kappa^r_M)=\{L\in L_g(R)$, for all $x\in h(M)$, $x\neq o$, and $a\in h(R)$, $(L:a)\not\subset\mathrm{Ann}_R x\}$. Moreover, we have that $\mathcal{L}(\kappa_{\underline{M}})\cap L_g(R)=\mathcal{L}(\kappa^r_M)$.

<u>Proof</u> : By definition of κ^r_M we have :

$$\mathcal{L}(\kappa^r_M)=\{L\in L_g(R),\mathrm{HOM}_R(R/L,\ E^g(M))=o\}\ .$$

Let $L \in \mathcal{L}(\kappa^r_M)$ and suppose there exist $x \neq o$ in $h(M)$ and $a \in h(R)$ such that $(L:a) \subset \mathrm{Ann}_R x$. Since $(L:a) \in \mathcal{L}(\kappa^r_M)$ it then follows that $\mathrm{Ann}_R x \in \mathcal{L}(\kappa^r_M)$. But this entails that there would exist a nonzero $f \in \mathrm{HOM}(R/\mathrm{Ann}_R x, E^g(M))$ what contradicts $R/\mathrm{Ann}_R x \in \mathcal{T}(\kappa^r_M)$. Conversely, supposing that $\mathrm{HOM}_R(R/L, E^g(M)) \neq o$ for some $L \in L_g(R)$ with the property that for all $x \neq o$ in $h(M)$ and $a \in h(R)$, $(L:a) \not\subset \mathrm{Ann}_R x$, we get that there exists a graded morphism $f : R/L \to E^g(M)$, $f \neq o$. Now since M is gr. essential in $E^g(M)$ there exists an $\bar{a} \in R/L$ with representative $a \in h(R)$ such that $x = f(\bar{a}) \neq o$ is an element of $h(M)$. Clearly $(L:a) \subset \mathrm{Ann}_R x$, contradiction. That $\mathcal{L}(\kappa_{\underline{M}})$ is as stated is well known. Further, the inclusion $\mathcal{L}(\kappa_{\underline{M}}) \cap L_g(R) \subset \mathcal{L}(\kappa^r_M)$ is obvious. For the converse inclusion consider $L \in \mathcal{L}(\kappa^r_M)$, $x \neq o$ in M and $a \in R$. Decompose x as $x_1 + \dots + x_m$, a as $a_1 + \dots + a_n$ and assume that these decompositions are so that we have ascending degrees. Since $x_m \neq o$ there can be found a $\lambda_1 \in h(R)$ such that $\lambda_1 a_1 \in L$ and $\lambda_1 x_m \neq o$. Similarly, we find $\lambda_2 \in h(R)$ such that $\lambda_2\lambda_1 a_2 \in L$ and $\lambda_2\lambda_1 x_m \neq o$. Using a recurrence argument we obtain an element $\lambda_n \in h(R)$ such that $\lambda_n \dots \lambda_2\lambda_1 a_n \in L$, $\lambda_n \dots \lambda_2\lambda_1 x_m \neq o$. Putting $\lambda = \lambda_n \dots \lambda_2 \lambda_1 \in h(R)$, we have that $\lambda x \neq o$ while $\lambda a \in L$, hence $(L:a) \not\subset \mathrm{Ann}_R x$, i.e. $L \in \mathcal{L}(\kappa_{\underline{M}})$. □

<u>12.1.7. Theorem.</u> Let R be a graded ring and let $(\mathcal{T},\mathcal{F})$ be a rigid torsion theory in R-gr. Let κ be the kernel functor on R-gr corresponding to $(\mathcal{T},\mathcal{F})$ and let $\mathcal{L}(\kappa)$ be its graded filter. Denote by $(\mathcal{T}_1,\mathcal{F}_1)$, respectively $(\mathcal{T}_2,\mathcal{F}_2)$, the hereditary torsion theory in R-mod generated by $\mathcal{T}$, respectively co-generated by $\mathcal{F}$. Let κ_1, $\mathcal{L}_1$, respectively κ_2, $\mathcal{L}_2$ be the corresponding kernel functors and filters. We have the following properties.

1.a. $\mathcal{L}_1 = \{L \text{ left ideal of } R,\ (L)_g \in \mathcal{L}(\kappa)\}$

b. If $M \in$ R-gr then $\kappa_1(\underline{M}) = \underline{\kappa(M)}$

c. $\mathcal{L}_1$ is the smallest filter in R containing $\mathcal{L}(\kappa)$ and $\mathcal{L}_1 \cap L_g(R) = \mathcal{L}(\kappa)$.

2.a. $\mathcal{L}_2 = \{L \text{ left ideal of } R,\ \mathrm{Hom}_R(R/L, E(\underline{M})) = o \text{ for all } M \in \mathcal{F}\}$

b. If $M \in$ R-gr then $\kappa_2(\underline{M}) = \underline{\kappa(M)}$

c. $\mathcal{L}_2 \cap L_g(R) = \mathcal{L}(\kappa)$.

Proof : 1.a. and then easily 1.c., is proved in a straightforward way. As for 1.b., it is clear that $\underline{\kappa(M)} \subset \kappa_1(\underline{M})$ because for a rigid kernel functor κ an element x is in $\kappa(M)$ if and only if its annihilator in R is in $\mathcal{L}(\kappa)$. Conversely, if $x \in \kappa_1(\underline{M})$ then $\mathrm{Ann}_R x \in \mathcal{L}_1$ i.e. $(\mathrm{Ann}_R x)_g \in \mathcal{L}(\kappa)$. Writing $x = x_1 + \dots + x_n$ with $\deg x_1 < \dots < \deg x_n$ we see that, for homogeneous $a \in (\mathrm{Ann}_R x)_g$ we have $ax_i = 0$, $i = 1, \dots, n$. Hence $x_i \in \kappa(M)$ and $x \in \kappa(M)$ follows. Note that we have herewith also shown that κ is induced on R-gr by κ_1 in the way described in Lemma 12.1.5. Therefore, all rigid kernel functors on R-gr may be considered as being induced on R-gr by some kernel functor on R-mod.

2.a. Is obvious since $(\mathcal{T}_2, \mathcal{F}_2)$ is co-generated by $\mathcal{F}$ in R-mod.

2.b. Since $M/\kappa(M) \in \mathcal{F}$, $\underline{M/\kappa(M)} \in \mathcal{F}_2$ follows, but as $\kappa_2(\underline{M})$ maps to a submodule of $\underline{M/\kappa(M)}$ which also has to be in $\mathcal{T}_2$, this means that $\kappa_2(\underline{M}) \subset \underline{\kappa(M)}$. If $F \in \mathcal{F}(\kappa)$ then from $\kappa_1(\underline{F}) = 0$ it follows that, $\kappa_1 \leq \kappa_{\underline{F}}$, hence $\kappa_1 \leq \inf\{\kappa_{\underline{F}}, F \in \mathcal{F}\} = \kappa_2$. Therefore, for every $M \in$ R-gr, $\kappa_1(\underline{M}) = \underline{\kappa(M)} \subset \kappa_2(\underline{M})$, thus $\underline{\kappa(M)} = \kappa_2(\underline{M})$.

2.c. Follows from 2.b. and the fact that the rigid kernel functor κ is uniquely determined by its graded filter.

By Lemma 12.1.6 we also have that $\kappa = \inf\{\kappa^r_M, M \in \mathcal{F}\}$ while $\kappa_2 = \inf\{\kappa_{\underline{M}}, M \in \mathcal{F}\}$.

12.1.8. Remarks. 1. It is clear that κ_1 is a graded kernel functor on R-mod whereas κ_2 is not a graded kernel functor.

2. The rigid torsion theories in R-gr form a set which may be partially ordered by putting $\kappa \leq \kappa'$ if and only if $\mathcal{L}(\kappa) \subset \mathcal{L}(\kappa')$. This ordering coincides with the ordering induced by the ordering on the kernel functors on R-gr, however for non-rigid functors this ordering cannot be checked on the graded filters alone.

3. The operation $\wedge$ restricted to rigid torsion theories coincides with the operation $\wedge$ induced in the set of rigid kernel functors by the $\wedge$ in R-mod. Here again we have that $\kappa = \bigwedge_i \kappa_i$ for rigid κ and κ_i is equivalent to $\mathcal{L}(\kappa) = \bigcap_i \mathcal{L}(\kappa_i)$ and again this criterion does not work for non-rigid kernel functors.

4. More about co-generated torsion theories will be in Section 12.3.

12.2. Examples of Torsion Theories in R-gr.

The singular submodule $Z(M)$ of $M \in R$-gr will be $\{x \in \underline{M},\ \mathrm{Ann}_R x$ is an essential left ideal$\}$ i.e. the singular submodule of $\underline{M}$ in R-mod.

12.2.1. Lemma. If $M \in R$-gr then $Z(M)$ is a graded submodule.

Proof : Pick an $x \neq o$ in $Z(M)$ and write $x = x_1 + \dots + x_n$ with $\deg x_1 < \dots < \deg x_n$, assuming that $x_j \neq o$ for all $j = 1, \dots, n$. Put $L = \mathrm{Ann}_R x$, then $L^{\sim} \subset \mathrm{Ann}_R x_n$. Since L is an essential left ideal, $L^{\sim}$ is essential and thus $\mathrm{Ann}_R x_n$ is an essential left ideal; thus $x_n \in Z(M)$. The lemma now follows by recurrence on n. □

It may easily be verified that Z defines a rigid preradical on R-gr. The associated radical Z_G satisfies $Z_G(M)/Z(M) = Z(M/Z(M))$. To a radical there may be associated an idempotent kernel functor (cf. [36]) in the usual way. The kernel functor K_G corresponding to the radical Z_G is rigid ; K_G may be called Goldie's graded kernel functor.

The dependence of $Z(-)$ on the module category under consideration is especially easy to deal with in the case of polynomial ring extension, as follows:

12.2.2. Proposition. Let R be an arbitrary ring, $M \in R$-mod. Then $Z_{R[X]}(M[X]) = Z_R(M)[X]$.

Proof : Take $m X^n \in Z_R(M)[X]_n$ with $m \neq o$. Since $m \in Z_R(M)$ it follows that $\mathrm{Ann}_R m$ is left essential in R, and this implies that $(\mathrm{Ann}_R m)[X]$ is left essential in $R[X]$. Now $\mathrm{Ann}_{R[X]} mX^n = (\mathrm{Ann}_R m)[X]$, thus $mX^n \in Z_{R[X]}(M[X])$, hence $Z_R(M)[X] \subset Z_{R[X]}(M[X])$. For the converse inclusion we only have to establish, on view of Lemma 12.2.1, that any homogeneous element of $Z_{R[X]}(M[X])$ is in $Z_R(M)[X]$. Let mX^n be in $Z_{R[X]}(M[X])$, then $\mathrm{Ann}_{R[X]} mX^n = (\mathrm{Ann}_R m)[X]$ and consequently $(\mathrm{Ann}_R m)[X]$ is left essential in $R[X]$, therefore $\mathrm{Ann}_R m$ is left essential in R i.e. $m \in Z_R(M)$ and $mX^n \in Z_R(M)[X]$ follows. □

The rigid torsion theory (T,F) co-generated in R-gr by R, i.e. $T = \{M \in R\text{-gr},\ \mathrm{HOM}_R(M, E^g(R)) = o\}$ where $E^g(R)$ is the injective hull of $\underline{R}$ in R-gr, is called Lambek's graded torsion theory. It's filter $\mathcal{L}$ is given by:

$$\mathcal{L} = \{L \in L_g(R),\ (L : a) \not\subset \mathrm{Ann}_R b \text{ for all } a,b \in h(R),\ b \neq o\}.$$

Lambek's torsion theory in R-mod is co-generated by (T,F). Theorem 12.1.7 implies that $L \in \mathcal{L}$ yields that $\underline{L}$ is dense in R.

Let S be a multiplicatively closed set in R. To S we associate a kernel functor $\underline{\kappa}_S$ on R-mod, which is given by its filter

$$\mathcal{L}(\underline{\kappa}_S) = \{\underline{L} \text{ left ideal of } R,\ (\underline{L}:r) \cap S \neq \emptyset \text{ for all } r \in R.\}.$$

12.2.3. Proposition. Let S be a multiplicatively closed system consisting of homogeneous elements and not containing zero then $\underline{\kappa}_S$ is a graded kernel functor in R-mod i.e. there corresponds a rigid kernel functor κ_S in R-gr to S, the filter of which is given by.

$$\mathcal{L}(\kappa_S) = \{L \in L_g(R), (L:r) \cap S \neq \emptyset \text{ for all } r \in h(R)\}.$$

Proof : Let $\underline{L} \in \mathcal{L}(\underline{\kappa}_S)$ i.e. $(\underline{L}:r) \cap S \neq \emptyset$ for all $r \in R$. If r is in h(R) this yields that $sr \in \underline{L}$ for some $s \in S$ i.e. $sr \in (\underline{L})_g$ and $((\underline{L})_g : r) \cap S \neq \emptyset$. If r is not homogeneous, write $r = r_1 + \dots + r_n$ with $\deg r_1 < \dots < \deg r_n$. Take $s_n \in (L:r_n) \cap S$, $s_{n-1} \in (L:s_n r_{n-1}) \cap S$, $\dots, s_1 \in (L:s_2 \dots s_n r_1) \cap S$, then, putting $s = s_1 s_2 \dots s_n$, we have $sr \in (\underline{L})_g$ with $s \in S$ i.e. $(\underline{L})_g \in \mathcal{L}(\underline{\kappa}_S)$ or $\underline{\kappa}_S$ is graded. The other statements follow directly from Lemma 12.1.5 and Theorem 12.1.7. □

12.2.4. Remark. If S satisfies the left Ore conditions in R then :
$\kappa_S(M) = \{m \in M,\ sm = o \text{ for some } s \in S\}$ for any $M \in$ R-gr.

Proof : Immediatly from 12.2.3. and 12.1.6. (1.b). □

Another example of a torsion theory in R-gr may be constructed as follows. Let (T,F) be the torsion theory in R-gr generated by the class of simple objects in R-gr. It is not hard to see that this is a rigid torsion theory and that an object

M of T is a semi-Artinian object of R-gr. i.e. the Gabriel dimension of $\underline{M}$ in R-mod is at most 1.

Recall that a torsion theory in R-mod is said to be symmetric if its associated filter allows a cofinal system of ideals of R. Similarly we define that a rigid kernel functor κ (or rigid torsion theory) in R-gr is symmetric if $\mathcal{L}(\kappa)$ has a cofinal set consisting of graded ideals.

12.2.5. Lemma. If κ is a rigid kernel functor on R-gr such that $\underline{\kappa}$ is symmetric on R-mod then κ is symmetric.

Proof : If $L \in \mathcal{L}(\kappa)$ then $\underline{L} \in \mathcal{L}(\underline{\kappa})$ and $\underline{L}$ contains an ideal I of R such that $I \in \mathcal{L}(\underline{\kappa})$. Since $\underline{\kappa}$ is a graded kernel functor on R-mod, $(\underline{I})_g \in \mathcal{L}(\underline{\kappa})$. Hence : $(I)_g \in \mathcal{L}(\kappa)$, $(I)_g \subset L$ and $(I)_g$ is a graded ideal of R. □

Now let R be a left Noetherian graded ring, I a graded ideal of R, P a prime ideal of R, and consider :

$$\mathcal{L}(I) = \{L \in L_g(R),\ L \supset I^n \text{ for some } n \in \mathbb{N}\}$$

$$\mathcal{L}(R-P) = \{L \in L_g(R),\ L \supset RsR \text{ for some } s \in R-P\}.$$

Then κ_I, associated to the filter $\mathcal{L}(I)$, and κ_{R-P}, associated to the filter $\mathcal{L}(R-P)$, are rigid idempotent kernel functors on R-gr; they are induced in R-gr by the functors $\underline{\kappa}_I$, $\underline{\kappa}_{R-P}$ on R-mod, which are given by their filters

$$\mathcal{L}(\underline{I}) = \{L \text{ left ideal of } R,\ L \supset \underline{I}^n \text{ for some } n \in \mathbb{N}\}$$

$$\mathcal{L}(R-P) = \{L \text{ left ideal of } R,\ L \supset RsR \text{ for some } s \in R-P\},$$

which have been constructed and used in [30], [31].

12.3. Injective Objects and Torsion Theories

If $M \in$ R-gr, let $E(\underline{M})$ be the injective hull of $\underline{M}$ in R-mod.

Let $M \in R$-gr be such that the singular radical $Z(M) = o$. Put : $E'_m = \{x \in E(\underline{M})$, there is a graded essential left ideal L such that $L_n x \subset M_{n+m}$ for all $n \in \mathbb{Z}$. It is easy enough to verify that $E' = \bigoplus_{m \in \mathbb{Z}} E'_m$ is a graded R-module containing M as a graded submodule.

12.3.1.. Lemma. Let $M \in R$-gr be such that $Z(M) = o$, then $\underline{E}'$ is the maximal R-submodule of $E(\underline{M})$ with the property that E' is graded and it contains M as a graded submodule.

Proof. Suppose that we have $M \subset N$ in R-gr such that $\underline{M} \subset \underline{N} \subset E(\underline{M})$. If $x \in N_m$ then $(\underline{M} : x)$ is a graded left ideal of R which is also essential. Since, $(\underline{M} : x)_n . x \subset M \cap N_{n+m} = M_{n+m}$, $x \in E'_m$ follows. ∎

12.3.2. Proposition. If $M \in R$-gr is such that $Z(M) = o$ then $E^g(M) \cong E'$ in R-gr.

Proof. In any Grothendieck category the injective hull may be characterized as being a maximal essential extension. By Lemma I.3.3.13., if $E^g(M)$ is a graded essential extension of M then it is an essential extension of $\underline{M}$ i.e. up to isomorphism we may assume : $\underline{E^g(M)} \subset E(\underline{M})$. The foregoing lemma now yields $E^g(M) \in E'$ i.e. $E' = E^g(M)$. ∎

Note that, even in case $Z(M) \neq o$, $h(E^g(M))$ consists of $x \in E(\underline{M})$ for which there is a graded essential left ideal L such that $L_n x \subset H_{n+m}$ for all $n \in \mathbb{Z}$ while $Lx \neq o$.

12.3.3. Remark. A graded module which is injective in R-mod is injective in R-gr. A kernel functor κ in R-gr is cogenerated by M if and only if it is cogenerated by $E^g(M)$ if and only if $\underline{\kappa}$ is cogenerated by $E(\underline{M})$ in R-mod.

A graded R-module M is said to be a graded supporting module for the kernel functor κ on R-gr if $\kappa(M) = o$ and $\kappa(M/N) = M/N$ for every graded submodule $N \neq o$. A kernel functor κ on R-gr is said to be graded prime if it is cogenerated by a graded supporting module. A kernel functor κ on R-gr is said to be projective if $R_+ \in \mathcal{L}(\kappa)$. The projective graded prime kernel functors on R-gr form a set which we denote R-Pr.

From now on we will write $\underline{\kappa}$ for the kernel functor on R-mod associated to κ on R-gr in the way described in Theorem 12.1.7 (i.e. the κ_1 there constructed is denoted by $\underline{\kappa}$). It is clear that if $M \in$ R-gr is such that $\underline{\kappa}_M$ is a prime kernel functor on R-mod with supporting module $\underline{M}$, then κ_M is a graded prime kernel functor on R-gr with graded supporting module M. Note that a graded kernel functor $\underline{\kappa}$ on R-mod such that the induced κ on R-gr is graded prime, need not be prime. For example take R to be $k[X]$, where k is a field, X a variable given positive degree, and consider $M = R \in$ R-gr. The kernel functors κ_X and $\underline{\kappa}_X$ on R-gr, resp. R-mod, may both be given by their filter which is generated by the set $\{(X),\dots,(X^n),\dots,k[X]\}$. It is obvious that $k[X]$ is a graded supporting module for κ_X but $\underline{k[X]}$ is not a supporting module for $\underline{\kappa}_X$. Warning : if M is graded then $\kappa_{\underline{M}}$ on R-mod need not be a graded kernel functor. However if $M \in$ R-gr is such that $\kappa_{\underline{M}}$ is a graded kernel functor then it follows from Lemma 12.1.6. that $\kappa_M = \underline{\kappa}^r_M$.

<u>12.3.4. Example</u>. Let R be a left Noetherian graded ring and P a graded prime ideal of R. It is well-known that the Lambek-Michler torsion theory with kernel functor $\kappa_{\underline{P}}$ on R-mod is co-generated by $E_R(R/P)$, hence also by $E^g_R(R/P)$. However $\kappa_{\underline{P}}$ is not necessarily a graded kernel functor; for example, take R to be a commutative integral domain, pick $s \in R-P$ and let s be non-homogeneous (this is possible if R is non-trivially graded), then Rs cannot contain an element of h(R-P) and $Rs \in \mathcal{L}(\kappa_{\underline{P}})$.

Clearly an $M \in$ R-gr of the form $\bigoplus_n E(-n)$ for some $E \in$ R-gr can never be a supporting module for a kernel functor κ on R-gr, therefore graded prime κ will in most cases not be rigid. Therefore if we say that κ on R-gr is a <u>rigid prime kernel functor</u> we mean to say that $\kappa = \bigwedge_n \kappa'(n)$ where κ' is a graded prime kernel functor, i.e. $\kappa' = \kappa^r_M$ for some M which is a graded supporting module for κ_M.

<u>12.3.5. Lemma</u>. If $M \in$ R-gr then κ^r_M is the largest rigid kernel functor on R-gr such that $\kappa^r_M(M) = o$.

<u>Proof</u> : If κ is a rigid kernel functor on R-gr such that $\kappa > \kappa^r_M$, $\kappa(M) = o$, then the

fact that the ordering of rigid functors may be checked by looking at their filters yields that there exists an $L \in \mathcal{L}(\kappa) - \mathcal{L}(\kappa_M^r)$ i.e. R/L is a graded module which is not κ_M^r-torsion, therefore by Lemma 12.1.6, there exists a nonzero graded morphism of degree p say, $f : R/L \to E^g(M)$. On the other hand R/L is κ-torsion and κ is rigid, therefore $f(R/L)$ is κ-torsion. But $\kappa(M) = o$ yields $\kappa(E^g(M)) = o$ hence $f(R/L) = o$, contradiction. Conversely if κ is any rigid kernel functor on R-gr such that $\kappa(M) = o$ then for every $I \in \mathcal{L}(\kappa)$, $HOM_R(R/I, M) = o$ follows from the rigidity of κ. By Lemma 12.1.6, $I \in \mathcal{L}(\kappa_M^r)$. So we have $\mathcal{L}(\kappa) \subset \mathcal{L}(\kappa_n^r)$ meaning that $\kappa \leqslant \kappa_M^r$ because both are rigid. □

12.3.6. Theorem. If $M \in$ R-gr has the property that $\kappa_{\underline{M}}$ is a prime kernel functor on R-mod with supporting module $\underline{M}$, then κ_M^r is a rigid prime kernel functor with supporting module M and κ_M is a graded prime kernel functor, $\mathcal{L}(\kappa_M^r) = \mathcal{L}(\kappa_{\underline{M}}) \cap L_g(R)$.

Proof : Immediately from Lemma 12.1.6. and the above. □

12.3.7. Corollary. If P is a graded prime ideal of R then we may correspond to P a rigid kernel functor on R-gr which may be induced on R-gr by means of a prime kernel functor $\kappa_{\underline{P}}$ on R-mod. Let us denote that rigid prime kernel functor associated to P by κ_P (normally we would have written κ_P^r but there will be no confusion.
If R is a left Noetherian and positively graded ring then κ_P is projective if and only if $P \in \mathrm{Proj}\ R = \{$graded prime ideals of R not containing $R_+\}$.

Proof : Consider the injective hull $E_R(R/P)$ of R/P in R-mod. It is well known that $E_R(R/P)$ is a supporting module for the torsion theory $\kappa_{\underline{P}}$ co-generated by it. Hence also $E_R^g(R/P)$ is a supporting module for $\kappa_{\underline{P}}$ and $\kappa_{\underline{P}}$ is co-generated by $E_R^g(R/P)$. By the foregoing theorem we may conclude that the rigid kernel functor κ_P which corresponds to the filter $\mathcal{L}(\kappa_{\underline{P}}) \cap L_g(R)$ is as stated in the corollary. If $P \in \mathrm{Proj}\ R$ then $R_+ + P/P = (R/P)_+$. If there exists a left ideal of R/P which does not intersect $(R/P)_+$ properly then there is an $x \neq o$ in R/P such that $(R/P)_+ x = o$, contradiction. Thus $(R/P)_+$ is essential in R/P and as such it contains a regular element i.e. $R_+ \cap G(P) \neq \emptyset$. Since R_+ is an ideal of R, $(R_+ : r) \cap G(P) \neq \emptyset$ for every $r \in R$ i.e. $R_+ \in L_g(R) \cap \mathcal{L}(\kappa_{\underline{P}})$.

Conversely, suppose that $R_+ \in \mathcal{L}(\kappa_P)$. Then it follows from Lemma 12.1.6. that for all $y \in h(E^g(R/P))$, $R_+ \not\subset \mathrm{Ann}_R y$, evidently $R_+ \not\subset P$ follows. □

We do not expound further the analogies between the correspondences, Spec $R \to$ Proj R and R-Sp $\to$ R-Pr, (where R-Sp denotes the set of prime kernel functors on R-mod).

Let us point out that combination of our results on Ass and on the injectives in R-gr allows to characterize graded fully bounded left Noetherian rings as follows.

12.3.8. Theorem. Let R be a positively graded left Noetherian ring. By $\mathcal{E}^g(R)$ we mean the equivalence classes of graded indecomposable injective objects for the relation : isomorphic up to suspension. The following statements are equivalent to one another :

1. The natural correspondence : $\mathrm{Spec}_g\, R \to \mathcal{E}^g(R)$ is a one-to-one correspondence.
2. Graded indecomposable injectives in R-gr are graded isotypic (i.e. E is gr. isotypic if $E = \bigoplus_i E_i$ with the E_i being the suspensions of some E_0).
3. For every $P \in \mathrm{Spec}_g R$ and for every left graded essential L in R/P we have that L contains an ideal I of R/P.
4. For every rigid kernel functor κ on R-gr, κ is symmetric.

Moreover if R is graded fully left bounded (i.e. one of the above conditions holds) then equivalently : R is gr. left Artinian or graded prime ideals of R are gr. maximal.

12.4. Graded Rings and Modules of Quotients.

The general theory of localization in a Grothendieck category may be applied so as to obtain the construction of objects of quotients with respect to some kernel functor or torsion theory. In the case of R-gr the interesting point of this is that for rigid torsion theories, the objects of quotients constructed, relate in a very nice way to the object of quotients constructed in R-mod. The exposition of this link between localization in R-mod and localization in R-gr is subject of this section.

Throughout κ will be an idempotent kernel functor on R-gr, $(\mathcal{T},\mathcal{F})$ will be the associated torsion theory in R-gr and $\mathcal{L}(\kappa)$ will be the filter of κ. Let $M \in$ R-gr. Define $Q_\kappa^g(M) = \varinjlim_{L \in \mathcal{L}} \mathrm{HOM}_R(L, M/\kappa(M))$; this graded R-module is said to be the object of quotients of M in R-gr with respect to κ.

One easily checks that $Q_\kappa^g(R)$ is, in a natural way, endowed with a graded ring structure, while $Q_\kappa^g(M)$ is a $Q_\kappa^g(R)$-module i.e. in $Q_\kappa^g(R)$-gr. It is clear that Q_κ^g is a left exact functor in R-gr which commutes with arbitrary suspensions.

For further use we state :

12.4.1. Lemma. Let R be a positively graded left Noetherian ring then R is an R_o-ring of finite type.

Proof : Let $x_1, \dots, x_m$ be homogeneous elements generating R_+ as a left ideal and let S be the ring generated by R_o and $\{x_1, \dots, x_m\}$. By construction $R_o \subset S$. We proceed by induction on j such that $R_j \subset S$. Suppose that $R_j \subset S$ for all $j < i \in \mathbb{N}$. An $x \in R_i$, $i > o$, can be written as $\sum_{t=1}^m r_o x_s$ with $r_s \in R_{i-d_s}$, $d_s = \deg(x_s)$. Since $d_s > o$ for all s, $r_s \in S$ for all s and $x \in S$ follows. Thus $R = S$. □

Let R be a positively graded ring throughout this section. Let $\underline{\kappa}$ be a graded kernel functor on R-mod and let $M \in$ R-gr. For each $m \in \mathbb{Z}$ put :

$$N_m = \{x \in Q_{\underline{\kappa}}(\underline{M}),\ \text{there is a graded left ideal } L \in \mathcal{L}(\underline{\kappa}) \text{ such that } L_n x \subset (M/\underline{\kappa}(\underline{M}))_{n+m} \text{ for all } n \in \mathbb{Z}\}\ .$$

(Note that $M/\underline{\kappa}(\underline{M})$ is graded because of 12.1.5).

The set N_m is well defined if m does not depend on the choice of $L \in \mathcal{L}(\underline{\kappa})$. However if $L_1 \in \mathcal{L}(\underline{\kappa})$ is graded and such that for some $x \neq o \in Q_{\underline{\kappa}}(\underline{M})$ we have :

$(L_1)_n x \subset (M/\underline{\kappa}(\underline{M}))_{n+r}$ for all $n \in \mathbb{Z}$, then $L \cap L_1 \in \mathcal{L}(\kappa)$ is graded and for some $n \in \mathbb{Z}$, $(I \cap J)_n\ x \neq o$ since $(I \cap J)x = o$ contradicts $o \neq x \in Q_{\underline{\kappa}}(\underline{M})$. So we obtain :

$o \neq (I \cap J)_n x \subset (M/\underline{\kappa}(\underline{M}))_{n+m} \cap (M/\underline{\kappa}(\underline{M}))_{n+v}$, entailing $m = v$.

Obviously $\bigoplus_{m \in \mathbb{Z}} N_m$ is a graded R-module which we will call the graded module of quotients of $\underline{M}$ at $\underline{\kappa}$ in R-mod, and denote it by $gQ_{\underline{\kappa}}(\underline{M})$, with these notations we have :

12.4.2. Proposition. $g\,\underline{Q}_\kappa(R)$ is a graded ring containing $R/\underline{\kappa}(R)$ as a graded subring. The graded ring structure of $g\,\underline{Q}_\kappa(R)$ is the unique ring structure compatible with its graded R-module structure. For every $M\in R$-gr, $g\,\underline{Q}_\kappa(M)$ is a graded $g\,\underline{Q}_\kappa(R)$-module.

Proof : It is immediate that $g\,\underline{Q}_\kappa(R)$ is a graded left R-module which is contained in $\underline{Q}_\kappa(R)$ and which contains $R/\underline{\kappa}(R)$ as a graded R-submodule. Pick $x,y\in g\,\underline{Q}_\kappa(R)$ and consider $xy\in\underline{Q}_\kappa(R)$ (the latter is a ring by general localization theory in R-mod). Without loss of generality one may take $x\in(g\underline{Q}_\kappa(R))_m$, $y\in(g\,\underline{Q}_\kappa(R))_n$ for some $n,m\in\mathbb{Z}$. Let $I,J\in\mathcal{L}(\underline{\kappa})$ be graded left ideals of R such that :

$$I_k x\subset(R/\underline{\kappa}(R))_{m+k} \text{ for all } k\in\mathbb{Z},$$

$$J_\ell y\subset(R/\underline{\kappa}(R))_{n+\ell} \text{ for all } \ell\in\mathbb{Z}.$$

If $(J:x)\cap I=o$ then $\underline{\kappa}$ is trivial and there is nothing to prove. If $L=(J:x)\cap I\neq o$ then we find $h\in\mathbb{Z}$ such that there is a nonzero $c\in L_h$. Hence $cxy\in I_{n+m}y\subset R_{h+m+n}$. Consequently $L_h xy\subset R_{h+m+n}$ for all $h\in\mathbb{Z}$, thus $xy\in(g\,\underline{Q}_\kappa(R))_{m+n}$. It follows that $g\,\underline{Q}_\kappa(R)$ is a graded ring and that this ring structure is determined by the graded R-module structure. Uniqueness of this ring structure as such follows from the fact that $g\,\underline{Q}_\kappa(R)$ is a subring of $\underline{Q}_\kappa(R)$, the ring structure of the latter being uniquely determined by its R-module structure. Similar argumentation shows that $g\,\underline{Q}_\kappa(M)$ is a graded $g\,\underline{Q}_\kappa(R)$-module. □

12.4.3. Theorem. Let $\underline{\kappa}$ be a graded kernel functor on R-mod, $M\in R$-gr. Then we have:

$$(g\,\underline{Q}_\kappa(M))_m=\varinjlim_{L\in\mathcal{L}(\underline{\kappa})\cap L_g(R)} HOM(L,M/\underline{\kappa}(M))_m\ .$$

Proof : It is well-known that $\underline{Q}_\kappa(M)=\varinjlim_{L\in\mathcal{L}(\underline{\kappa})}\mathrm{Hom}(L,M/\underline{\kappa}(M))$. By construction, $x\in(g\,\underline{Q}_\kappa(M))_m$ if and only if x is represented by a graded morphism of degree m : $m_x : L\to M/\underline{\kappa}(M)$ for some $L\in\mathcal{L}(\underline{\kappa})\cap Lg(R)$. Since $\mathcal{L}(\underline{\kappa})\cap Lg(R)$ is cofinal in $\mathcal{L}(\underline{\kappa})$ the theorem follows easily. □

12.4.4. Corollary. Let κ be a rigid kernel functor on R-gr. Then $\underline{\kappa}$ is the graded kernel functor on R-mod the torsion theory of which is generated by the torsion theory of κ. By Theorem 12.1.7. $\underline{\kappa}(\underline{M}) = \kappa(M)$ and $\mathcal{L}(\underline{\kappa}) \cap L_g(R) = \mathcal{L}(\kappa)$. By definition of $Q^g_\kappa(M)$ and by Theorem 12.4.3. it follows that $Q^g_\kappa(M) = gQ_{\underline{\kappa}}(\underline{M})$. Thus the functors Q^g_κ and $gQ_{\underline{\kappa}}$ are naturally equivalent functors on R-gr. Because of this we will use the notation $Q^g_\kappa(M)$ for the object of quotients in R-gr associated to the graded kernel functor $\underline{\kappa}$ on R-mod.

<u>12.4.5. Proposition.</u> Q^g_κ is a covariant left exact endofunctor in R-gr : Moreover Q^g_κ is functorial with respect to graded morphisms of arbitrary degree.

<u>Proof</u> : We only have to prove the last statement. Let $N, M \in$ R-gr, $f : N \to M$ a graded morphism of degree m. Then $Q_{\underline{\kappa}}(f) : Q_{\underline{\kappa}}(N) \to Q_{\underline{\kappa}}(M)$ is R-linear. If $x \in (Q^g_\kappa(N))_k$ then there is an $I \in \mathcal{L}(\kappa)$ such that $I_n x \subset (N/\underline{\kappa}(\underline{N}))_{n+k}$ for all $n \in \mathbb{N}$. Hence for all $n \in \mathbb{N}$ we have $I_n h(x) \subset h((N/\underline{\kappa}(N))_{n+k})$ and the latter is exactly $f((N/\kappa(N))_{n+k})$ hence contained in $(M/\kappa(N))_{n+k+m}$ where the graded morphism of degree $m : N/\kappa(N) \to M/\kappa(M)$, deriving from f, has again been denoted by f. Therefore $h(x) \in (Q^g_\kappa(M))_{k+m}$ and this states exactly that $h|Q^g_\kappa(N)$ is a graded morphism of degree m. □

Recall the following definitions. A kernel functor $\underline{\kappa}$ on R-mod is said to be a <u>finite type</u> if for every $L \in \mathcal{L}(\underline{\kappa})$ there exists an $L' \in \mathcal{L}(\underline{\kappa})$, $L' \subset L$ and L' is finitely generated. We say that κ has <u>property T</u> if one of the following equivalent conditions is fulfilled :

i) $Q_{\underline{\kappa}}$ is exact and commutes with direct sums.

ii) For all $M \in$ R-mod $Q_{\underline{\kappa}}(M) = Q_{\underline{\kappa}}(R) \otimes_R M$.

iii) For all $L \in \mathcal{L}(\underline{\kappa}) : Q_{\underline{\kappa}}(R) = Q_{\underline{\kappa}}(R)\, j_{\underline{\kappa}}(L)$ where $j_{\underline{\kappa}}$ is the canonical morphism $R \to R/\underline{\kappa}(R)$.

iv) The quotient category with respect to $\underline{\kappa}$ coincides with $Q_{\underline{\kappa}}(R)$-mod.

v) Every $Q_{\underline{\kappa}}(R)$-module is $\underline{\kappa}$-torsion free as an R-module.

From iii) one easily derives that a kernel functor having property T has finite type.

<u>12.4.6. Lemma.</u> Let κ be the rigid kernel functor on R-gr induced by the graded kernel functor $\underline{\kappa}$ on R-mod. If $\underline{\kappa}$ has finite type then κ has finite type.

Proof : Easy. □

12.4.7. Proposition. Let κ be the rigid kernel functor on R-gr induced by the graded kernel-functor $\underline{\kappa}$ on R-mod and suppose that $\underline{\kappa}$ has finite type, then $Q^g_\kappa(M) = Q_{\underline{\kappa}}(M)$ for all $M \in$ R-gr.

Proof : Since $Q^g_\kappa(M) = \varinjlim_{L \in \mathcal{L}(\kappa)} \mathrm{HOM}_R(L, M/\kappa(M))$, it follows from Lemma I.3.3.2; combined with the fact that the finitely generated graded left ideals of $\mathcal{L}(\kappa)$ form a cofinal system, that $Q^g_\kappa(M) = \varinjlim_{L \in \mathcal{L}(\underline{\kappa})} \mathrm{Hom}_R(L, M/\underline{\kappa}(M))$ (note $\underline{\kappa}(M) = \kappa(M)$). □

12.4.8. Corollary. If R is a left Noetherian positively graded ring and $\underline{\kappa}$ a graded kernel functor on R-mod inducing κ on R-gr then $Q_{\underline{\kappa}}(M)$ is graded for every $M \in$ R-gr, and $Q_{\underline{\kappa}}(M) = Q^g_\kappa(M)$.

If $\underline{\kappa}$ is a projective kernel functor on R-mod then let S_m be the set of $x \in Q_{\underline{\kappa}}(R)$ such that there exists a graded left ideal $L \in \mathcal{L}(\underline{\kappa})$ such that $Lx \subset j_\kappa(R)$ and $L_n x \subset j_\kappa(R)_{n+m}$ for all $n \geq n_0$ for some fixed $n_0 \in \mathbb{N}$, and consider $S = \oplus_m S_m$.

12.4.9. Lemma. If $\underline{\kappa}$ is projective then S is a graded ring.

Proof : We only have to check that m in the definition of S_m does not depend on the choice of $L \in \mathcal{L}(\underline{\kappa})$; the rest of the proof is then formally the same as the proof of Proposition 12.4.2. So let $J \in \mathcal{L}(\kappa)$ be such that $Jx \subset j_\kappa(R)$ and $J_n x \subset (j_\kappa(R))_{n+m'}$ for all $n \geq n'_0$, some fixed $n'_0 \in \mathbb{N}$.
For all $n \geq \max\{n_0, n'_0\} = N_0$ we have then that :
$H = J \cap L \in \mathcal{L}(\kappa)$ is such that $H_n x \subset (j_\kappa(R))_{n+m} \cap (j_\kappa(R))_{n+m'}$. Thus if $m \neq m'$ then $H_n x = o$ for all $n \geq N_0$, consequently $R_{\geq N_0} Hx = o$ follows then from the fact that the gradation of R is positive. However $(R_+)^{N_0} \subset R_{\geq N_0}$ and $(R_+)^{N_0} \in \mathcal{L}(\kappa)$ since the fact that κ is idempotent implies that $\mathcal{L}(\kappa)$ is closed under taking products, thus $R_{\geq N_0} \in \mathcal{L}(\kappa)$ and thus $R_{\geq N_0} H \in \mathcal{L}(\kappa)$ or $x \in \underline{\kappa}\ Q_\kappa(R) = o$. □

12.4.10. Proposition. Let $\underline{\kappa}$ be a projective kernel functor then the ring S is nothing but $Q^g_\kappa(R)$.

Proof : The definitions imply that $Q^g_\kappa(R) = g\underline{Q}_\kappa(R) \subset S$. Take $x \in S_m$ to be nonzero i.e. let $L \in \mathcal{L}(\kappa)$ be such that $Lx \subset j_\kappa(R)$ and $L_n x \subset (j_\kappa(R))_{m+n}$ for all $n \geq n_o$, for some fixed $n_o \in \mathbb{N}$. Now, if $L_k x = o$ for all $k < n_o$ then $L_n x \subset (j_\kappa(R))_{n+m}$ holds for all $n \in \mathbb{N}$ and thus $x \in Q^g_\kappa(R)_m$ follows. Suppose that $L_k x \neq o$ for some $k < n_o$ and fix this k. Pick a nonzero $y \in L_k x$ and let $z \in j_\kappa^{-1}(y)$. Assume first that $j_\kappa(R_n)L_k x = o$ for all $n \geq n_o - k$. Then $R_n z$ maps to zero in $j_\kappa(R)$ for all $n \geq n_o - k$, consequently $R_{\geq n_o - k}\, z$ maps to zero in $j_\kappa(R)$ i.e. $R_{\geq n_o - k}\, z \subset \underline{\kappa}(R)$. However $R_{\geq n_o - k} \in \mathcal{L}(\underline{\kappa})$ since $(R_+)^{n_o - k} \subset R_{\geq n_o - k}$, thus $z \in \underline{\kappa}(R)$ and $y = o$, contradiction. Therefore we may assume that $j_\kappa(R_n)L_k x \neq o$ for <u>some</u> $n \geq n_o - k$. We obtain : $o \neq j_\kappa(R_n)L_k x \subset L_{n+k} x \subset (j_\kappa(R))_{n+k+m}$ since $k+n \geq n_o$. From this it is evident that $L_k x \subset j_\kappa(R)_{m+k}$, hence in all cases : $L_n x \subset (j_\kappa(R))_{m+n}$ for all $n \in \mathbb{N}$, i.e. $x \in (Q^g_\kappa(R))_m$.

12.4.11. Remark. In a similar way it follows that the projective module of quotients of $M \in R\text{-gr}$ at a projective kernel functor $\underline{\kappa}$ is nothing but $Q^g_\kappa(M)$.

12.5. More Properties of Graded Localization.

First let us expound some results about graded morphisms and their localizations, which for graded morphisms of degree o are trivial or covered by general localization theory in R-gr. R is positively graded throughout and κ is rigid unless otherwise stated.

12.5.1. Lemma. Let κ be a kernel functor on R-gr, let $M \in R\text{-gr}$, $L \in \mathcal{L}(\kappa)$. A graded morphism of arbitrary degree : $f : L \to M$ extends in a unique way to a graded morphism of the same degree $h : R \to Q^g_\kappa(M)$.

Proof : By faithful $\underline{\kappa}$-injectivity of $Q_{\underline{\kappa}}(M)$ we know that f extends to an R-linear map $\bar{f} : R \to Q_{\underline{\kappa}}(M)$. It is clear that in proving the lemma we may suppose that M is κ-torsion free. Now $\bar{f}(L_n) = L_n \bar{f}(1)$ and $\bar{f}(L_n) = f(L_n) \subset M_{n+m}$ so for all $n \in \mathbb{N}$: $L_n \bar{f}(1) \subset M_{n+m}$ i.e. $\bar{f}(1) \in Q^g_\kappa(M)_m$ where m is the degree of f. Now since R is graded $\bar{f}$ maps R to $Q^g_\kappa(M)$ and it is obviously the unique extension of f.

<u>12.5.2. Proposition.</u> Let $M,N,S\in R\text{-gr}$ be such that $N\subset M$ and $\kappa(M/N)=M/N$ for some kernel functor κ on R-gr. Any graded morphism f of degree d, $f:N\to Q^g_\kappa(S)$ extends in a unique way to a graded morphism h of degree d, $h:M\to Q^g_\kappa(S)$.

<u>Proof.</u> Let $x\in M$ and let $I\in\mathcal{L}(\kappa)$ be such that $Ix\subset N$. The definition of $Q^g_\kappa(M)$ entails that $\underline{Q}_\kappa(\underline{Q^g_\kappa(M)})=\underline{Q}_\kappa(\underline{M})$ and therefore $Q^g_\kappa(Q^g_\kappa(M))=Q^g_\kappa(M)$. Let x be homogeneous and apply the foregoing lemma to m_x, i.e. right multiplication by x, we get the following commutative diagram of graded morphisms

$$\begin{array}{ccc} I & \xrightarrow{\quad m_x \quad} & N \\ \downarrow & & \downarrow \\ R & \xrightarrow{\quad h_x \quad} & Q^g_\kappa(N). \end{array}$$

The graded morphism h_x composes with $Q^g_\kappa(f):Q^g_\kappa(N)\to Q^g_\kappa(S)$ and yields a graded morphism $\psi_x:R\to Q^g_\kappa(S)$. Define $h(x)=\psi_x(1)$. It is straightforward to verify that ψ_x does not depend on the choice of $I\in\mathcal{L}(\kappa)$ and that the map h is graded morphism of degree equal to deg f which satisfies all requirements. □

Recall that for $\underline{M}\in R\text{-mod}$, $\underline{\kappa}$ a kernel functor on R-mod, we can characterize $\underline{Q}_\kappa(\underline{M})$ by the exact sequence in R-mod :

$$0\to\underline{M}/\underline{\kappa}(\underline{M})\longrightarrow\underline{Q}_\kappa(\underline{M})\xrightarrow{\ \pi(\underline{M})\ }\kappa(E(\underline{M})/\underline{M})\longrightarrow 0$$

where $E(\underline{M})$ is an injective hull of $\underline{M}$ in R-mod.

If $M\in R\text{-gr}$ then we have the following exact diagram in R-mod :

$$\begin{array}{ccccccccc} 0 & \longrightarrow & \underline{M} & \longrightarrow & E(\underline{M}) & \xrightarrow{\ \pi(\underline{M})\ } & E(\underline{M})/\underline{M} & \longrightarrow & 0 \\ & & \| & & \uparrow & & \uparrow & & \\ 0 & \longrightarrow & \underline{M} & \longrightarrow & \underline{E^g(M)} & \longrightarrow & \underline{E^g(M)}/\underline{M} & \longrightarrow & 0 \end{array}$$

and the bottom row may be considered as an exact row in R-gr.

12.5.3. Proposition. Let $M \in R\text{-gr}$ and let κ be a kernel functor in R-gr. Then $Q^g_\kappa(M)$ is characterized by the following exact sequence in R-gr : (let M' be $M/\kappa(M)$):

$$0 \longrightarrow M' \longrightarrow Q^g_\kappa(M) \xrightarrow{\pi^g} \kappa(E^g(M')/M') \longrightarrow 0$$

where π^g is the restriction of the canonical map $E^g(M') \to E^g(M')/M'$.

Proof : The fact that $Q^g_\kappa(M) = g\, Q_{\underline{\kappa}}(M)$ guarantees that it is a graded submodule of $E^g(M')$, such that $Q^g_\kappa(M)/M'$ is κ-torsion. Suppose $x \in E^g(M')$ and $Ix \subset M'$, i.e. $I_n x \subset M'$ for all $n \in \mathbb{Z}$, for some $I \in \mathcal{L}(\kappa)$. Put $x = \sum_\ell ' x_\ell$ and $i \in I_n$, then $\sum_\ell ' i x_\ell = \sum_j ' m_j$ for some $m_j \in M'$. Since M' is a graded submodule of $E^g(M')$ it follows that $ix_\ell = m_{n+\ell}$ for all ℓ, n and $i \in I_n$. Consequently $I_n x_\ell \subset M'_{n+\ell}$, thus $x_\ell \in Q^g(M)_\ell$ and $x \in Q^g_\kappa(M)$ follows. □

12.5.4. Corollary. Let κ be a kernel functor on R-gr, let $\underline{\kappa}$ be the cogenerated kernel functor on R-mod, then, for every $M \in R\text{-gr}$, $Q^g_\kappa(M)$ is the largest graded module containing $M/\kappa(M)$ as a submodule, which is contained in $Q_{\underline{\kappa}}(\underline{M})$.

$Q^g_\kappa(R)$-modules may be considered as being R-modules by restriction of scalars with respect to the canonical graded ring morphism $R \to Q^g_\kappa(R)$.

12.5.5. Lemma. Let $\underline{M} \in Q^g_\kappa(R)\text{-mod}$, $N \in R\text{-gr}$. If a map $f : \underline{M} \to Q^g_\kappa(N)$ is R-linear then it is also $Q^g_\kappa(R)$-linear.

Proof : Pick $m \in M$, $\lambda \in Q^g_\kappa(R)$. For some $L \in \mathcal{L}(\kappa)$ we have $L\lambda \subset R/\kappa(R)$ and thus $Lf(\lambda x) = f(L\lambda x) = L\lambda . f(x)$. Hence $f(\lambda x) - \lambda f(x) \in \kappa(Q^g_\kappa(N)) = o$. □

12.5.6. Theorem. Let κ be a rigid kernel functor on R-gr and let $\underline{\kappa}$ be the kernel functor on R-mod associated to it, then the following statements are equivalent :

1. $Q^g_\kappa(R) j_\kappa(L) = Q^g_\kappa(R)$ for all $L \in \mathcal{L}(\kappa)$, where j_κ is the canonical morphism $R \to R/\kappa(R)$.
2. If $\underline{M} \in Q^g_\kappa(R)\text{-mod}$. then $\underline{\kappa}(\underline{M}) = o$.
3. If $M \in Q^g_\kappa(R)\text{-gr}$. then $\kappa(M) = o$.
4. $Q^g_\kappa(-) = Q^g_\kappa(R) \otimes_R -$ (natural equivalence in R-gr).
5. $\underline{\kappa}$ has property T.

Proof : 1⇒2. If $x \in \underline{\kappa}(\underline{M})$ then $Lx = o$ for some $L \in \mathcal{L}(\kappa)$, hence $Q_\kappa^g(R)(Lx) = o$. By restriction of scalars this means that $Q_\kappa^g(R) j_\kappa(L) x = o$ i.e. $x = o$.

2⇒3. Obvious.

3⇒4. Let $M \in R$-gr. The canonical R-linear map $\underline{M} \to Q_{\underline{\kappa}}(\underline{M})$ factorizes as

$$\underline{M} \xrightarrow{\alpha} Q_{\underline{\kappa}}(R) \underset{R}{\otimes} \underline{M} \xrightarrow{\beta} Q_{\underline{\kappa}}(\underline{M}) \ .$$

Since $Q^g(M)$ is a $Q^g(R)$-module we may restrict this sequence to a sequence of graded morphisms :

$$M \xrightarrow{\alpha^g} Q^g(R) \underset{R}{\otimes} M \xrightarrow{\beta^g} Q^g(M) \ ,$$

(Note : $\alpha(m) = 1 \otimes m$, $\beta(\sum_i q_i \otimes m_i) = \sum q_i m_i$ and similar for α^g and β^g).

By 3., $\kappa(Q_\kappa^g(R) \underset{R}{\otimes} M) = o$ hence Ker $\alpha^g = \kappa(M)$ that Ker $\alpha^g \subset \kappa(M)$ is clear since Ker $\alpha \subset \kappa(\underline{M})$) thus Ker $\alpha = \underline{\kappa}(\underline{M})$. Then β and a fortiori β^g has to be a monomorphism. Because of the foregoing lemma β^g is $Q_\kappa^g(R)$-linear, hence Imβ^g is a graded $Q_\kappa^g(R)$-submodule of $Q_\kappa^g(M)$. Note that β^g is a graded morphism of degree o. Since Imβ^g contains $M/\kappa(M)$ it follows that $Q_\kappa^g(M)/\text{Im}\beta^g$ is κ-torsion and a graded $Q_\kappa^g(R)$-module, thus by 3. again, Imβ^g - $Q_\kappa^g(M)$ or β^g is an isomorphism.

4⇒1. We have $Q_\kappa^g(R) = Q_\kappa^g(L) = \beta^g(Q_\kappa^g(R) \underset{R}{\otimes} L) = Q_\kappa^g(R) j_\kappa(L)$ for every $L \in \mathcal{L}(\kappa)$.

1⇒5. Since $Q_\kappa^g(R)$ is a subring of $Q_{\underline{\kappa}}(R)$ and since $\mathcal{L}(\underline{\kappa})$ has a cofinal system of graded left ideals it follows from 1. that $Q_{\underline{\kappa}}(R) j_\kappa(L) = Q_\kappa(R)$ for every $L \in \mathcal{L}(\kappa)$; i.e. $\underline{\kappa}$ has property T.

5⇒1. If $\underline{\kappa}$ has property T then it has finite type, thus Proposition 12.4.7. yields that $\underline{Q_\kappa^g(M)} = Q_{\underline{\kappa}}(\underline{M})$ for all $M \in R$-gr. In particular for $L \in \mathcal{L}(\kappa)$ we have that $Q_\kappa^g(R) = Q_\kappa^g(L) = Q_{\underline{\kappa}}(\underline{L}) = Q_{\underline{\kappa}}(R) j_\kappa(\underline{L}) = Q_\kappa^g(R) j_\kappa(L)$. □

12.5.7. Definition. A rigid kernel functor κ in R-gr is said to have property T if Q_κ^g is exact and commutes with direct sums.

12.5.8. Theorem. Let κ be a rigid kernel functor on R-gr and let $\underline{\kappa}$ be the associated kernel functor on R-mod. Equivalently :

1. $\underline{\kappa}$ has property T in R-mod.

2. κ has property T in R-gr.

Proof : 1 ⇒ 2. If $\underline{\kappa}$ has property T then it has finite type i.e. Q_κ^g and $Q_{\underline{\kappa}}$ are equivalent functors in R-gr. Since $Q_{\underline{\kappa}}$ is exact and commutes with direct sums in R-mod it will certainly by true that Q_κ^g is exact and commutes with direct sums in R-gr.
2 ⇒ 1. Take $L \in \mathcal{L}(\kappa)$, let j_κ be the canonical morphism $R \to R/\kappa(R)$ and consider $Q_\kappa^g(R)j_\kappa(L)$. This is a graded $Q_\kappa^g(R)$-module thus it may be obtained as the quotient of a free graded $Q_\kappa^g(R)$-module which is then isomorphic to $F = Q_\kappa^g(R)(n_1) \oplus \dots \oplus Q_\kappa^g(R)(n_t) \oplus \dots$ for $n_1, \dots, n_t, \dots \in \mathbf{Z}$. Since Q_κ^g commutes with suspensions and direct sums we see that $F = Q_\kappa^g(R(n_1) \oplus \dots \oplus R(n_t) \oplus \dots)$, hence $Q_\kappa^g(F) = F$. Furthermore the epimorphism $\pi : F \to Q_\kappa^g(R)j_\kappa(L)$ yields by exactness of Q_κ^g an epimorphism :

$$\pi' : F = Q_\kappa^g(F) \to Q_\kappa^g(Q_\kappa^g(R)j_\kappa(L))$$

which is the unique extension of π. Thus, since $F = Q_\kappa^g(F)$, $\pi' = \pi$ and $Q_\kappa^g(Q_\kappa^g(R)j_\kappa(L)) = Q_\kappa^g(R)j_\kappa(L)$ follows. On the other hand $Q_\kappa^g(L)/j_\kappa(L)$ is κ-torsion hence $Q_\kappa^g(L) = Q_\kappa^g(R)$ yields that $Q_\kappa^g(R)/Q_\kappa^g(R)j_\kappa(L)$ is κ-torsion too i.e. $Q_\kappa^g(Q_\kappa^g(R)j_\kappa(L)) = Q_\kappa^g(R)$. Combining these results yields that for every $L \in \mathcal{L}(\kappa)$, we have that $Q_\kappa^g(R)j_\kappa(L) = Q_\kappa^g(R)$. Theorem 12.5.6, 1 ⇔ 5. then conludes the proof. □

12.5.9. Lemma. Let κ be a rigid kernel functor on R-gr and let $\underline{M}, \underline{M}' \in Q_\kappa^g(R)$-mod. Suppose that $f : \underline{M} \to \underline{M}'$ is an R-homomorphism such that $Q_\kappa^g(R)f(\underline{M})$ is $\underline{\kappa}$-torsion free as an R-module, then f is $Q_\kappa^g(R)$-linear.

Proof : Pick $x \in \underline{M}$, $\lambda \in Q_\kappa^g(R)$ and let $L \in \mathcal{L}(\kappa)$ be such that $L\lambda \subset j_\kappa(R)$. Then $f(L\lambda x) = Lf(\lambda x) = L\lambda . f(x)$, hence $f(\lambda x) - \lambda f(x) \in \kappa(Q_\kappa^g(R)f(\underline{M})) = 0$. □

12.5.10. Lemma. Let $\underline{M} \in Q_\kappa^g(R)$-mod be $\underline{\kappa}$-torsion free when considered as an R-module, then the following conditions are equivalent :

1. $\underline{M}$ is injective in R-mod.

2. $\underline{M}$ is injective in $Q_\kappa^g(R)$-mod.

Proof : 1 ⇒ 2. If $\underline{E}$ is the injective hull of $\underline{M}$ in $Q^g_\kappa(R)$-mod then

$$o \to \underline{M} \to \underline{E} \to \underline{E}/\underline{M} \to o$$

splits in R-mod, i.e. there is an R-linear $f : \underline{E} \to \underline{M}$ such that $f|\underline{M} = 1_{\underline{M}}$. Since $\underline{E}$ is $\underline{\kappa}$-torsion free we may apply Lemma 12.5.9. and conclude that f is $Q^g_\kappa(R)$-linear, meaning that the sequence also splits in $Q^g_\kappa(R)$-mod. So $\underline{E} = \underline{M}$ follows.

2 ⇒ 1. Let $\underline{E}$ be the injective hull of M in R-mod. Since $\underline{\kappa}(\underline{M}) = o$ we have $\underline{\kappa}(\underline{E}) = o$ hence $Q_{\underline{\kappa}}(\underline{E}) = \underline{E}$ and $\underline{E}$ is a $Q_{\underline{\kappa}}(R)$-module thus, restricting scalars, also a $Q^g_\kappa(R)$-module. By the assumption 2, the sequence

$$o \to \underline{M} \to \underline{E} \to \underline{E}/\underline{M} \to o$$

splits in $Q^g_\kappa(R)$-mod. But then $\underline{M}$ is a direct summand of $\underline{E}$ in R-mod too, thus $\underline{E} = \underline{M}$. □

12.5.11. Corollary. Let κ be a rigid kernel functor on R-gr, $M \in$ R-gr, and suppose that $\kappa(M) = o$ and that $\underline{M}$ is also a $Q^g_\kappa(R)$-module. Then : $E^g_R(M) \cong E^g_{Q^g_\kappa(R)}(M)$ in R-gr..

The foregoing lemmas may be applied in the study of compatibility of kernel functors as in [33], cf.loc.cit. for general theory concerning reflectors and Giraud subcategories of Grothendieck categories. Again the fact that $Q^g_\kappa(R)$ is not necessarily a generator of the quotient category $G(\kappa)$ associated to κ in R-gr, cannot prevent us from describing the induced rigid kernel functors in $G(\kappa)$ by their filters in $Q^g_\kappa(R)$.

All kernel functors κ on R-gr considered will be rigid, $\underline{\kappa}$ will be the associated kernel functor on R-mod. Let κ and κ_1 be kernel functors on R-gr, then $\underline{\kappa_1}$ induces a kernel functor $\underline{\nu}_1$ in $Q^g_\kappa(R)$-mod by taking the $\underline{\kappa}_1$-torsion $Q^g_\kappa(R)$-modules for the torsion class of $\underline{\nu}_1$. It is easily checked that $\underline{\nu}_1$ is a graded kernel functor inducing a rigid kernel functor ν_1 on $Q^g_\kappa(R)$-gr. We say that $\underline{\kappa}_1$ is Q^g_κ-compatible if $\underline{\kappa_1}\,\underline{Q^g_\kappa} = \underline{Q^g_\kappa}\,\underline{\kappa_1}$ in R-gr, or equivalently $\kappa_1 Q^g_\kappa = Q^g_\kappa \kappa_1$ in R-gr; in this case we also say that $\underline{\kappa_1 \text{ is } Q^g_\kappa\text{-compatible}}$. It is not hard to verify that, if κ_1 is Q^g_κ-compatible, then $\mathcal{L}(\nu_1)$ has a cofinal subset consisting of the $Q^g_\kappa(L)$, $L \in \mathcal{L}(\kappa_1)$, so if κ has property T (then $\underline{\kappa}$ has property T!) it follows that $\mathcal{L}(\nu_1) = \{Q^g_\kappa(R) j_\kappa(L),\ L \in \mathcal{L}(\kappa_1)\}$.

<u>12.5.12. Example</u>. If $\kappa_1 \geqslant \kappa$ then κ_1 is Q^g_κ-compatible.

<u>12.5.13. Proposition</u>. Let κ_1, κ be rigid kernel functors on R-gr such that κ_1 is Q^g_κ-compatible, then we have that $Q^g_{\kappa_1} Q^g_\kappa(M) \cong Q^g_{\nu_1} Q^g_\kappa(M)$ for all $M \in$ R-gr.

<u>Proof</u> : By definition : $\nu_1 Q^g_\kappa(M) = Q^g_\kappa(\nu_1(M))$. Put : $\varepsilon(-) = E^g_{Q^g_\kappa(R)}(-)$,

$N = Q^g_\kappa(M)/\nu_1 Q^g_\kappa(M) = Q^g_\kappa(M)/Q^g_\kappa(\kappa_1(M))$. (*)

By left exactness of Q^g_κ, $N \to Q^g_\kappa(M/\kappa_1(M))$ is monomorphic, hence N is κ-torsion free and in $Q^g_\kappa(R)$-mod. Corollary 12.5.11 entails that $\varepsilon(N) = E^g_R(N)$. Furthermore we have the following isomorphisms in R-gr :

$$\nu_1(\varepsilon(N)/N) \cong \kappa_1(\varepsilon(N)/N) \cong \kappa_1(E^g_R(N)/N).$$

Applying Proposition 12.5.3. we obtain the following commutative diagram in R-gr :

$$\begin{array}{ccccccccc} 0 & \longrightarrow & N & \longrightarrow & Q^g_{\nu_1}(N) & \longrightarrow & \nu_1(\varepsilon(N)/N) & \longrightarrow & 0 \\ & & = \big| & & \big\downarrow \gamma & & \big\downarrow \cong & & \\ 0 & \longrightarrow & N & \longrightarrow & Q^g_{\kappa_1}(N) & \longrightarrow & \kappa_1(E^g(N)/N) & \to & 0 \end{array}$$

where γ is the restriction of the graded isomorphism $\varepsilon(N) \to E^g_R(N)$, and where rows are exact. It follows from this that : $Q^g_{\nu_1}(N) \cong Q^g_{\kappa_1}(N)$ in R-gr. However, from (*) it is then clear that on one hand $Q^g_{\nu_1}(N) = Q^g_{\nu_1}(Q^g_\kappa(M))$ while on the other hand $Q^g_{\nu_1}(N) \cong Q^g_{\kappa_1}(N) = Q^g_{\kappa_1}(Q^g_\kappa(M))$. Lemma 12.5.9. yields that these isomorphisms are graded isomorphisms in $Q^g_\kappa(R)$-gr! □

Since κ has property T if and only if $\underline{\kappa}$ has property T the whole theory of compatibility as expounded in [33] may now easily be adapted to the graded case, in particular compatibility of symmetric graded kernel functors, as in [33], may be developped in a similar way (with an eye to projective geometry over a noncommutative ring).

II.13. GRADED PRIME IDEALS AND THE ORE CONDITION.

In this section R will always be a positively graded and left Noetherian ring. To a graded prime ideal P of R we associate a graded prime kernel functor κ_P as in 12.3.7 and a rigid graded symmetric kernel functor κ_{R-P} as in 12.2. For a left Noetherian ring R, to every kernel functor $\underline{\kappa}$ on R-mod there corresponds a symmetric kernel functor $\overset{\circ}{\underline{\kappa}}$ on R-mod which is the largest symmetric kernel functor smaller than κ.

13.1. Proposition. Let $\underline{\kappa}$ be a graded kernel functor on R-mod inducing the rigid kernel functor κ on R-gr, then $\overset{\circ}{\underline{\kappa}}$ induces a graded symmetric $\overset{\circ}{\kappa}$ on R-gr which is the largest (in the ordering of rigid kernel functors on R-gr) rigid symmetric kernel functor smaller than κ.

Proof : Follows in a straightforward way from the fact that an ideal of R which contains a graded left ideal of R also contains a graded ideal of R. □

For completeness sake we include some results on kernel functors associated to m-systems (semi-multiplicative systems) i.e. a set D such that for each pair $x,y \in D$ there exists a $z \in R$ such that $xzy \in D$. To an m-system D in a left Noetherian ring we may associate a symmetric kernel functor $\underline{\kappa}$ given by $\mathcal{L}(\underline{\kappa}) = \{L$ left ideal of R, $L \supset RdR$, for some $d \in D.\}$.

13.2. Theorem. Let R be a left Noetherian graded ring and let D be an m-system in R such that R-D is additionally closed then h(D) is an m-system and the symmetric kernel functor $\underline{\kappa}$ associated to h(D) is symmetric graded.

Proof : Pick $x \in h(D)$, $y \in h(D)$. Since D is an m-system there is a $z \in R$ such that $xzy \in D$. Write $z = \sum_{i=1}^{n} z_i$. If $z_i \neq o$ then $xz_iy \in R-D$ for all $z_i \neq o$ appearing in the expression for z, then $xzy \in R-D$, contradiction. Thus $xz_iy \in h(D)$ for some i and h(D) is an m-system in R. Obviously, the kernel functor associated to h(D) is symmetric graded on R-gr. □

If $\underline{\kappa}$ is a kernel functor on R-mod then $(\underline{\kappa})_g$ will be the rigid kernel functor on R-gr given by the filter : $\mathcal{L}((\underline{\kappa})_g) = \mathcal{L}(\underline{\kappa}) \cap L_g(R)$.

13.3. Lemma. Let P be a graded prime ideal of R and consider the following kernel functors on R-mod :

$\kappa_{\underline{P}}$ associated to the multiplicatively closed set G(P).

$\underline{\kappa}_{h(P)}$ associated to the multiplicatively closed set h(G(P)).

$\kappa_{\underline{R-P}}$ associated to the m-system R-P in R.

$\underline{\kappa}_{h(R-P)}$ associated to the m-system h(R-P) in R.

Then the following relations hold :

1. $(\kappa_{\underline{P}})_g \geq (\underline{\kappa}_{h(P)})_g = \kappa_{h(P)}$

2. $(\kappa_{\underline{R-P}})_g \geq (\underline{\kappa}_{h(R-P)})_g = \kappa_{h(R-P)}$

3. $\overset{\circ}{((\kappa_{\underline{P}})_g)} = (\kappa_{\underline{R-P}})_g$.

Proof : Trivial. □

13.4. Theorem. Let R be a positively graded left Noetherian graded ring and let $P \in \mathrm{Proj}\ R$ then $\kappa_P = \kappa_{h(P)}$ and $\kappa_{R-P} = \kappa_{h(R-P)}$, (where κ_P and κ_{R-P} are defined as in 12.3.7 and 12.2. resp).

Proof : Since κ_P is induced on R-gr by $\kappa_{\underline{P}}$ (cf. 12.3.7.) we have that $\mathcal{L}((\kappa_{\underline{P}})_g) = \mathcal{L}(\kappa_P)$ i.e. $\kappa_P = (\kappa_{\underline{P}})_g \geq \kappa_{h(P)}$. Now let $L \in \mathcal{L}(\kappa_P)$, let $\pi : R \to R/P$ be the canonical epimorphism. From $L \in \mathcal{L}(\kappa_P)$ it follows that $\pi(L)$ is a graded left ideal of R/P containing a left regular element, hence, that $\pi(L)$ is a graded left essential left ideal of the prime left Noetherian and positively graded ring R/P, with $(R/P)_+ \neq o$.

First we prove that if J is a graded left ideal of R/P such that h(J) consists of nilpotent elements then J = o. Note that if Ra has this property for some $a \in h(R)$ then aR has this property too, so it will be sufficient to prove that no right ideal J of R/P has that property! So suppose that J is a right ideal of R/P such that h(J) consists of nilpotent elements. If $a \neq o$ in h(J) then $Ra \neq o$,

i.e. for all $a \neq o$ in h(J), $\mathrm{Ann}_R^{\ell}(a) \neq R/P$.

The Noetherian hypothesis allows to choose a $b \in h(J)$, $b \neq o$, such that $\mathrm{Ann}_R^{\ell} b$ is

maximal amongst left annihilators of elements of $h(J)$ which are nonzero. Take $r_k \in R_k$ then $\mathrm{Ann}^{\ell}_R b = \mathrm{Ann}^{\ell}_R br_k$ but $br_k \in h(J)$ hence $(br_k)^t = o$ for some t such that $(br_k)^{t-1} \neq o$. Thus $\mathrm{Ann}^{\ell}_R b = \mathrm{Ann}^{\ell}_R (br_k)^{t-1}$, but then $br_k b = o$. Hence $bRb = o$, contradiction.
So we have shown that if J is a nonzero left ideal of R/P then I contains a non-nilpotent homogeneous element. Since $(R/P)_+ \neq o$ and $\pi(L)$ is essential in R/P, it follows that there exists a non-nilpotent homogeneous element a_1 in $\pi(L) \cap (R/P)_+$. Now repeat the argumentation of the proof of Proposition 9.2.3. to conclude that $\pi(L)$ contains a homogeneous regular element of positive degree q, say c. There is thus an element of degree q in R, d say, such that $d+p \in L$ for some $p \in P$ and $\pi(d+p) = c$. Since L is graded $d+p_q \in L$, and still $\pi(d+p_q) = c$ (since $p_q \in P$). Moreover $d+p_q$ is homogeneous of degree q and $d+p_q \in G(P)$ if and only if $d \in G(P)$; hence $d+p_q \in h(G(P))$ because $d \in G(P)$ follows from the fact that $\pi(d)$ is left- (hence right-) regular in R/P. Therefore $\kappa_P = \kappa_{h(P)}$ now follows from statement 1 in Lemma 13.3. By Proposition 13.1, $\overset{\circ}{\kappa}_{h(P)} = \overset{\circ}{\kappa}_P$ is induced in R-gr by $\overset{\circ}{\kappa}_{\underline{P}} = \kappa_{\underline{R-P}}$ hence $\overset{\circ}{\kappa}_{h(P)} = (\kappa_{\underline{R-P}})_g \geq \kappa_{h(R-P)}$ whereas $\kappa_{h(R-P)} \geq \overset{\circ}{\kappa}_{h(P)}$ is obvious from $h(G(P)) \subset h(R-P)$. The fact that $(\kappa_{\underline{R-P}})_g$ is exactly κ_{R-P} as defined in 12.2 finishes the proof of the second statement. □

<u>13.5. Remark.</u> A special case of this Theorem has been proved in [30] for prime ideals P of R which have the property that, for all $x,y \in R-P$, $xRy \cap yRx \cap (R-P) \neq \emptyset$. Exactly the extra information drawn from the graded version of Goldie's theorems made the complete solution to the problem given here possible.

<u>13.6. Proposition.</u> Let R be a left Noetherian positively graded ring and $P \in \mathrm{Proj}\, R$ a graded prime ideal such that the left Ore conditions with respect to G(P) hold, then the left Ore conditions also hold with respect to h(G(P)).

<u>Proof</u> : Given $s \in h(G(P))$, $r \in R$, consider $(Rs : r)$. Write $r = r_1 + \ldots + r_n$ with $\deg r_1 < \ldots < \deg r_n$ and put $L = \bigcap_{i=1}^{n} (Rs : r_i) \subset (Rs : r)$. Clearly L is a graded left ideal which is in $\mathcal{L}(\kappa_{\underline{P}})$ because of the left Ore condition with respect to G(P). Now $L \in \mathcal{L}((\kappa_{\underline{P}})_g) = \mathcal{L}(\kappa_P)$ and it follows from Theorem 13.5 that $L \in \mathcal{L}(\kappa_{h(P)})$, consequently there exists $s_1 \in h(G(P))$, $r_1 \in R$ such that $s_1 r = r_1 s$. □

13.7. Remark. The analogues of Theorem 13.4 and Proposition 13.6 hold in the non Noetherian case for those $P \in \mathrm{Proj}\, R$ such that R/P is a graded Goldie ring. (R is of course still positively graded!).

II.14. THE PRESHEAVES ON Proj R.

In this section R is again always a positively graded and left Noetherian ring. First we deduce a representation theorem for finitely generated modules over Proj R.

14.1. Theorem. Let $M \in R\text{-gr}$ be a finitely generated object then $Q_{R-P}(M) = o$ for all $P \in \mathrm{Proj}\, R$ if and only if $Q_{R_+}(M) = o$ if and only if there is an $n \in \mathbb{N}$ such that $M_m = o$ for all $m > n$.

Proof : By Lemma 12.4.1, R is generated as an R_o-ring by the finite number of chosen homogeneous generators for R_+ i.e. $R = R_o < r_1, \ldots, r_m >$ with $\deg r_m \geq \ldots \geq \deg r_1 > o$. If $P \in \mathrm{Proj}\, R$ then $Q_{R-P}(M) = o$ means that $I_P M = o$ for some graded ideal $I_P \not\subset P$. Put $I = \sum_{P \in \mathrm{Proj}\, R} I_P$ then $\mathrm{rad}\, I = P_1 \cap \ldots \cap P_r$ with $P_i \supset R_+$ for all $i = 1, \ldots r$. Consequently $R_+^N \subset I$ for some $N \in \mathbb{N}$. Hence $M = \kappa_{R_+}(M)$ or $Q_{R_+}(M) = o$. Conversely $M = \kappa_{R_+}(M)$ yields that $Q_{R-P}(M) = o$ for all $P \in \mathrm{Proj}\, R$ since $R_+ \not\subset P$.
Now let M be generated by the homogeneous elements $m_1, \ldots, m_t$ with $\deg m_1 \leq \deg m_2 \leq \ldots \leq \deg m_t$. If $M_m = o$ for all $m > n$ then $R_+^N M = o$ for $N > n \deg m_1$, hence $Q_{R_+}(M) = o$. Conversely, suppose that $R_+^N M = o$ for some $N \in \mathbb{N}$. If $N_o > N.\deg r_m$ then $R = R_o < r_1, \ldots, r_m >$ entails that $R_{N_o} \subset R_+^N$, thus for all $N_o > N.\deg r_m : R_{N_o} M = o$. Take $N_1 > \deg m_t + N.\deg r_m$ and $N_1 > N \deg r_m$ and consider $x \in M_{N_1}$ i.e. $x = x_1 m_1 + \ldots + x_t m_t$ with $\deg x_j > N.\deg r_m = N_o$ for all j. But then $x_j m_k \in R_{N_o} M = o$ for all j,k i.e. $x = o$, thus $M_m = o$ for all $m > N_1$. □

The following result is fundamental in the study of the projective spectrum :

14.2. Theorem. Fix $n_o \in \mathbb{N}$. For each $n \geq n_o$ let there be given an additive subgroup p_n of R_n. The following statements are equivalent :

1. There is a unique $P \in \mathrm{Proj}\, R$ such that $P \cap R_n = p_n$ for all $n \geq n_o$.
2.a. For all n,k and for all $m \geq n_o$ we have $R_n p_m R_k \subset p_{n+m+k}$.

b. Let $r \in R_n$, $t \in R_m$ and suppose that $n,m \geq n_o$ are such that $rR_k t \subset p_{n+m+k}$ for all k, then $r \in p_n$ or $t \in p_m$.

c. $p_n \neq R_n$ for some $n \geq n_o$.

Proof : 1 ⇒ 2. It is trivial that a. and b. are fulfilled if 1. holds. Now suppose that $p_n = R_n$ for all $n \geq n_o$. Then for all $k \neq o$ and all $a \in R_k$ we have that :

$$a R_{\ell_1} a \dots a R_{\ell_r} a \in R_{k(r+1) + \sum_i \ell_i} ,$$

for all $(\ell_1, \dots, \ell_r) \in \mathbb{N}^r$. Choose r such that $k(r+1) \geq n_o$, then the foregoing comes down to : $a R_{\ell_1} a \dots a R_{\ell_r} a \subset P$ for all $(\ell_1, \dots, \ell_r) \in \mathbb{N}^r$, hence $a \in P$ or $a \in P_k$, contradicting $R_+ \not\subset P$.

2 ⇒ 1. From c it follows that we may fix $n \geq n_o$ and an element $a \in R_n - p_n$. For each $m \in \mathbb{N}$ put :

$$P'_m = \{x \in R_m, \text{ for large } t, \text{ for all } (\ell_1, \dots, \ell_t) \in \mathbb{N}^t, \ a R_{\ell_1} a \dots a R_{\ell_t} x \subset p_{m+tn+\sum_i \ell_i}$$

Obviously, if $m \geq n_o$ then $P'_m = p_m$ because of b. For any $m \in \mathbb{N}$, P'_m is additive. Put $P = \Sigma P'_m$. Clearly P is an ideal of R because of a. and $P_n \neq R_n$ since $a \notin P_n$. Suppose $s \in R_\lambda$, $w \in R_\mu$ are such that $sR_k w \in P'_{\lambda+\mu+k}$ for all $k \in \mathbb{N}$. Since R is left Noetherian RsRw is generated by a finite number of homogeneous elements which may obviously be chosen to be of the form $sr_k w$. Thus we may choose $t \neq o$ large enough such that : for all $(\ell_1, \dots, \ell_t) \in \mathbb{N}^t$, for all chosen generators $sr_k w$ we have :

$$a R_{\ell_1} a \dots a R_{\ell_t} sr_k w \subset p_{\lambda+\mu+k+tn+\sum_i \ell_i}$$

For any t' larger than t (fixed as above) we obtain :

$$((RaR)^{t'})_\gamma sR_k w \subset p_{\gamma+\lambda+\mu+k} ,$$

for all k,γ. Let us write A = RaR, then for all $k, \gamma \in \mathbb{N}$:

$$(A^{t'})_\gamma \, s \, (A^{t'})_k \, w \subset p_{\gamma+\lambda+\mu+k}.$$

Now, $(A^{t'})_\gamma \neq o$ implies that $\gamma \geq nt'$ because deg $a = n > o$. Further : $A^{t'} \neq o$, because otherwise $a \in p_n$ would result from b. Again from b. it follows now that either, $(A^{t'})_\gamma$ s is in $p_{\gamma+\lambda}$ or $(A^{t'})_k$ w is in $p_{k+\mu}$ where both $\gamma+\mu$ and $k+\mu$ may be taken to be larger than $t'n > n_o$. The construction of P yields that either s or w must be in P and therefore we have established : $P \in \mathrm{Proj}\ R$. Assume that $Q \in \mathrm{Proj}\ R$ is different from P and suppose Q satisfies the conditions of 2.

Pick $b \in Q-P$ and suppose that b is homogeneous. If deg $b > o$ then for all $m \geq n_o$ and all $(\ell_1, \ldots, \ell_m) \in \mathbb{N}^m$ we have :

$$b\, R_{\ell_1}\, b \ldots b\, R_{\ell_{m-1}}\, b \subset Q \cap R_\lambda \subset p_\lambda \subset P$$

where $\lambda = md + \sum_i \ell_i$. So $(Rb)^m \subset P$, but this contradicts $b \notin P$. □

<u>14.3. Remark.</u> The above theorem is true if R is commutative, even in the absence of the left Noetherian condition. We will prove a similar theorem for special but non left Noetherian rings further but it is not known to us whether the above theorem holds for arbitrary positively graded rings in general.

<u>14.4. Proposition.</u> Let I be a graded ideal of R and let C be the set {J, J an ideal of R maximal such that $J \notin \mathcal{L}(I)$}. Put $gC = \{(J)_g, J \in C\}$. Then gC consists of graded prime ideals and $\mathcal{L}(I) = \bigcap_{P \in gC} \mathcal{L}(\kappa_{R-P})$. Moreover if $I \subset R_+$ then the elements of gC are in Proj R.

<u>Proof</u> : It is straightforward to check that C consists of prime ideals (see also [32]), thus gC consists of graded prime ideals. Obviously an ideal $H \in \mathcal{L}(I)$ if and only if $H \not\subset P$ for all $P \in C$. Since the filters we have to compare are both graded filters it follows that a filter basis for $\mathcal{L}(I)$ is obtained by taking the graded ideals not contained in any of the $(P)_g \in gC$. This proves that $\mathcal{L}(I) = \bigcap_{P \in gC} \mathcal{L}(\kappa_{R-P})$. If $I \subset R_+$ then for any $P \in gC$ we have that $P \not\supset I$ and a fortiori $P \not\supset R_+$, whence $P \in \mathrm{Proj}\ R$. □

<u>14.5. Proposition.</u> Let I be a graded ideal of R, then : $R_+ \cap \mathrm{rad}(I_+) = R_+ \cap \mathrm{rad}(I)$.

<u>Proof</u> : If P is a graded prime ideal containing I_+ then either $P \supset I$ or $P \supset R_+$. Note

that this property does hold if R is not necessarily left Noetherian. □

<u>14.6. Corollary.</u> Put $V_+(I) = \{P \in \operatorname{Proj} R,\ P \supset I\}$. Then :

$$V_+(I) = V_+(I_+) = V_+(\operatorname{rad}(I)) = V_+((\operatorname{rad}(I))_+) = V_+(\operatorname{rad}(I_+)) \quad .$$

Let κ_I be the symmetric graded kernel functor associated to $\mathcal{L}(I)$, Q_I^g the graded localization functor on R-gr corresponding to κ_I, $j_I : R \to Q_I^g(R)$ the corresponding canonical graded ring homomorphism of degree o. If $I \subset R_+$ then κ_I is a projective kernel functor on R-gr, because of Lemma 12.4.9 then $Q_I^g(M)$ for $M \in$ R-gr may also be given by $(Q_I^g(M))_m = \{x \in Q_I(M)$, there is a $J \in \mathcal{L}(I)$ such that $Jx \subset j_I(R)$ and $J_n x \subset (j_I(R))_{n+m}$ for all $n \geq n_0\}$.
Note that if $I \subset R_+$ then $J \in \mathcal{L}(I)$ if and only if $J_+ \in \mathcal{L}(I)$. An easy strengthening of Proposition 14.5 yields that : $R_+ \cap \operatorname{rad}(I) = R_+ \cap \operatorname{rad}(\bigoplus_{n \geq n_0} I_n)$ for each graded ideal I of R and each natural number n_0.

Endow Proj R with the topology induced by the Zariski topology of Spec R as follows. Put, for any ideal I of R : $V_+(I) = \{P \in \operatorname{Proj} R,\ P \supset I\}$, $X_+(I) = X - V_+(I)$ where $X = \operatorname{Proj} R$. In these definitions we may replace I by the smallest graded ideal of R containing I, so from now on, when we write $V_+(I)$ or $X_+(I)$, I is understood to be graded. By Corollary 14.6. we may assume further that $I \subset R_+$ and $V_+(I) = V_+(\bigoplus_{n \geq n_0} I_n)$ for each $n_0 \in \mathbb{N}$. Again from Corollary 14.6 it follows that $V_+(I)$ remains unaltered under taking radicals and positively graded parts of I in any order. The following relations are easily checked :

$$V_+(I+J) = V_+(I) \cap V_+(J)$$

$$V_+(IJ) = V_+(I \cap J) = V_+(I) \cup V_+(J) \quad .$$

This shows that the sets $X_+(I)$, I varying in the set of graded ideals (or those contained in R_+) of R, exhausts the open sets of the topology induced in X.
To an open set $X_+(I)$ the projective kernel functor R_I is associated i.e.

$\mathcal{L}(I_+) = \mathcal{L}(\kappa_{I_+}) = \{L \in L_g(\ell),\ L \in L_g(\ell),$ L contains an ideal J in $L_g(R)$ such that $\mathrm{rad}(J) \supset I_+\}$.

14.7. Lemma. For any open set $X_+(I)$ of Proj R we have :

$$\mathcal{L}(I_+) = \cap\{\mathcal{L}(\kappa_{R-P}), P \in X_+(I)\}.$$

Proof : If $P' \in X_+(I)$ then $P' \notin \mathcal{L}(\kappa_{I_+})$ hence $P' \subset P$ for some $P \in gC(I_+)$. Thus $\mathcal{L}(\kappa_{R-P}) \subset \mathcal{L}(\kappa_{R-P'})$. Applying Proposition 14.4 yields :

$$\mathcal{L}(\kappa_{I_+}) = \bigcap_{P \in gC(I_+)} \mathcal{L}(\kappa_{R-P}) = \bigcap_{P' \in X_+(I)} \mathcal{L}(\kappa_{R-P'}). \quad \square$$

It is easily verified that in case R is left Noetherian, $X_+(I)$ is quasi-compact in X. The following proposition may be verified by mimicing, step by step, the proof of the corresponding properties for Spec R mentioned in [32], taking care to use the up till now established graded theory where necessary :

14.8. Proposition. Assigning $Q^g_{I_+}(R)$ to $X_+(I)$ for each graded ideal I of R; defines a presheaf of graded rings $\underline{Q}^g(R)$: Open $(X)^{\mathrm{opp}} \to$ R-gr, over the Zariski topology on Proj R = X. If R is a left Noetherian graded ring then every open set is quasi-compact and the presheaf $\underline{Q}^g(R)$ is separated.

If $\underline{Q}^g(R)$ is separated then it may be embedded in a sheaf of rings which is obtained by applying the well known sheafification functor L (sheaf reflector) which associates to a separated presheaf P a sheaf LP which on $U \in \mathrm{Open}(X)$ is given by :

$$L\ (U) = \varprojlim_{\mathcal{U} \in \mathrm{Cov}(U)} \ \varinjlim_{V \in \mathcal{U}} P(V) \ ,$$

where Cov(U) is the poset of coverings of U by open sets of X. Since $\varprojlim$ and $\varinjlim$ are inner in R-gr and also in the category of graded rings (with graded morphisms of degree o) it follows immediately that $L\underline{Q}^g(R)$ is a sheaf of graded rings. For $M \in$ R-gr we obtain a presheaf $\underline{Q}^g(M)$ and a sheaf $L\underline{Q}^g(M)$ of graded R-modules. The following theorem is again an easily verified graded version of a similar theorem in [32].

<u>14.9. Theorem.</u> Let R be a positively graded left Noetherian prime ring then $LQ^g(R) = Q^g(R)$ i.e. $Q^g(R)$ is a sheaf.

<u>14.10. Remark.</u> The graded ring of "functions" for $Q^g(R)$ is defined to be : $\varinjlim_{X_+(I)} Q^g_I(R)$. This ring may be obtained by localizing R at $\kappa = V\{\kappa_{I_+}, I$ graded ideal of R$\}$, where the supremum (V) is defined as being the kernel functor induced on R-gr by $V\underline{\kappa}_{I_+}$ on R-mod.

<u>14.11. Theorem.</u> Let $P \in \mathrm{Proj}\ R$ then we have :

1. $\kappa_{R-P} = V\{\kappa_{I_+}, I_+$ such that $P \in X_+(I)\}$
2. The stalk of the sheaf $LQ^g(R)$ at P is exactly $Q^g_{R-P}(R)$.

<u>Proof</u> : 1. $P \in X_+(I)$ if and only if $I_+ \not\subset P$, thus $\kappa_I \leq \kappa_{R-P}$ for every graded ideal I such that $P \in X_+(I)$. Conversely, if $J \in \mathcal{L}(\kappa_{R-P})$ then J contains a nonzero graded ideal I such that $I_+ \not\subset P$, henco $P \in X_+(I)$ and also $J \in \mathcal{L}(\kappa_I)$.

2. The stalk of $LQ^g(R)$ at $P \in \mathrm{Proj}\ R$ is defined to be $\varinjlim_{P \in X_+(I)} Q^g_I(R) = S$. Since R is left Noetherian $\kappa_{R-P}(R)$ is finitely generated hence $J\,\kappa_{R-P}(R) = o$ for some graded ideal $J \not\subset P$, consequently $Q^g_J(R) \to Q^g_{R-P}(R)$ is a monomorphism. This remark entails that in the directed system considered, the maps $f_I : Q^g_I(R) \to Q^g_{R-P}$, $P \in X_+(I)$, eventually turn out to be monomorphisms for small I. Therefore we obtain a monomorphism of degree o, $f : S \to Q^g_{R-P}(R)$. An element $x \in (Q^g_{R-P}(R))_m$ may be represented as a graded morphism of degree m, $m_x : I \to R$ for some $I \in \mathcal{L}(\kappa_{R-P})$. Now m_x also represents an element y_I of $Q^g_I(R)$. By construction of the f_I it is clear that $f_I(y_I) = x$, therefore the image of y_I in S, y say, is such that $f(y) = x$ and this states exactly that f is an isomorphism of degree o (note that both the gradation of S and $Q^g_{R-P}(\ell)$ extend the gradation of R, so the bijective morphism of degree o is an isomorphism.

If X is any topological space, $\mathcal{R}$ a (pre-) sheaf of graded rings defined over X. Define a presheaf $\mathcal{R}_o$ by $\mathcal{R}_o(U) = \mathcal{R}(U)_o$ for each $U \in \mathrm{Open}(X)$, then.

14.12. Lemma. $\mathcal{R}_0$ is a sub (pre-) sheaf of $\mathcal{R}$.

Proof : Let ρ^U_V denote the restriction morphism $\mathcal{R}(U) \to \mathcal{R}(V)$ with respect to open sets $V \subset U$ in X. Since ρ^U_V is graded of degree o, the restriction of ρ^U_V to $R(U)_0$ maps $R(U)_0$ into $\mathcal{R}(V)_0$. As a subpresheaf of $\mathcal{R}$, $\mathcal{R}_0$ is separated. Further if $U = \{U_i\}_i$ is a covering of $U \in \mathrm{Open}(X)$ and if $f_i \in \mathcal{R}(U_i)$ are elements of degree o such that :
$\rho^{U_i}_{U_i \cap U_j}(f_i) = \rho^{U_j}_{U_i \cap U_j}(f_j)$, for all i,j;
then there is an $f \in \mathcal{R}(U)$ such that $\rho^U_{U_i}(f) = f_i$ for all i. Again, the fact that ρ^U_V has degree o implies $f \in \mathcal{R}(U)_0$, consequently $\mathcal{R}_0$ is a sheaf.

The sheaf of rings $(\underline{LQ}^g(R))_0$ defined over Proj R is called the structure sheaf of Proj R; it will be denoted simply by Proj. Since direct limits of graded morphisms of degree o "respect" taking homogeneous parts of degree o we find that : $\mathrm{Proj}(X_+(I)) = (Q^g_I(R))_0$ for every graded ideal I of R. $\mathrm{Proj}_P = (Q^g_{R-P}(R))_0$ for every $P \in \mathrm{Proj}\ R$.

In the commutative case graded prime ideals can be related to common prime ideals in some of the rings appearing in the structure sheaf of Proj. This relation is fully expressed by saying that Proj is a scheme i.e. Proj R has a covering by open sets $X_+(I)$, I in some set of graded ideals β, such that : $\mathrm{Proj}|X_+(I) \cong \mathrm{Spec}(Q^g_I(R))_0$; indeed it suffices to take for β the set of graded ideals generated by a single homogeneous element of R. The general non-commutative case looks to be difficult and it is still non-solved up to now. However in the sequel we show that this property does hold for the class of Zariski central rings, introduced in [33]. We need some ring theoretic introduction about these rings.

II. 15. GRADED ZARISKI CENTRAL RINGS.

15.1. Basic Facts About Zariski Central Rings.

Zariski central rings, studied in [33], are a special case of the

birational algebras considered in [31]. In this section we repeat some definitions and fundamental properties, referring to [31] and [33] for more details.
In this section R is a ring with unit and C denotes the centre of R. The ring R is said to be Zariski central if the Zariski topology of Spec $R = X$ has a basis of open sets $X(S) = \{P \in X, P \not\supset S\}$ where S varies through the subsets of C; equivalently : R is Zariski central if $\mathrm{rad}(I) = \mathrm{rad}(R(I \cap C))$ for every ideal of R. Examples of Zariski central rings are, semisimple Artinian rings, Azumaya algebras, rings $R = A[X,\varphi,\delta]$ where A is a simple ring and where the centre of R is not a field : in particular $A[X,\varphi]$ is Zariski central.

15.1.1. Lemma. If R is a Zariski central ring then the filters :

$$\mathcal{L}(I) = \{L \text{ left ideal of } R,\ L \supset J \text{ ideal of } R,\ \mathrm{rad}(J) \supset I\}$$

$$\mathcal{L}(R\text{-}P) = \{L \text{ left ideal of } R,\ L \supset J \text{ ideal of } R,\ J \not\subset P\},$$

are idempotent filters for every ideal I of R and every prime ideal P of R.

Because of the above lemma we are able to localize at κ_I, κ_{R-P}, the kernel functors on R-mod associated to $\mathcal{L}(I)$, $\mathcal{L}(R\text{-}P)$ resp., even if R is not necessarily left Noetherian. If κ is a symmetric kernel functor on R-mod then let $\mathcal{L}(\kappa^C)$ be the filter generated by the ideals I of C such that $RI \in \mathcal{L}(\kappa)$. The following statements are equivalent, for some $M \in R$-mod :

1. M is κ-torsion,
2. ${}_C M$ is κ^C-torsion.

The class of Zariski central rings is closed under taking epimorphic images.

15.1.2. Theorem. Let κ be a symmetric kernel functor on R-mod, R a Zariski central ring, and suppose that κ^C has property T then κ has property T. Moreover :

1. $Q_\kappa(R)$ is a central extension of R and $Q_\kappa(R)$ and $Q_{\kappa^C}(R)$ are isomorphic rings.
2. For every $M \in R$-mod the C-modules $Q_\kappa(M)$ and $Q_{\kappa^C}({}_C M)$ are isomorphic.
3. If I is an ideal of R then $Q_\kappa(R)\, j_\kappa(I)$ is an ideal of $Q_\kappa(R)$, where $j_\kappa : R \to Q_\kappa(R)$ is the canonical morphism.

4. There is a one-to-one correspondence between prime ideals of $Q_\kappa(R)$ and prime ideals of R which are not in $\mathcal{L}(\kappa)$.

15.1.3. Lemma. If D is a graded division ring then D is a Zariski central ring.

Proof. $D = D_0[X,X^{-1},\varphi]$ for some skewfield D_0 and an automorphism φ of D_0, cf. Theorem 6.3, unless $D = D_0$ is a skewfield and in the latter case the statement is trivial. Now $S = D_0[X,\varphi]$ is Zariski central, cf. [31] and localizing S at the Ore set $\{1,X,X^2,\ldots\}$ yields that D is Zariski central too. □

15.2. GZ-and ZG-Rings.

A positively graded ring R is said to be a GZ-ring if it is a Zariski central ring. A ring R, positively graded, is said to be a ZG ring if for every graded ideal I of R we have that $\mathrm{rad}(I) = \mathrm{rad}(R(I \cap C))$. A GZ-ring is a ZG-ring; we will deal with GZ-rings here, however, most results may also be obtained for the larger class of rings.

15.2.1. Theorem. Let R be a positively graded Zariski central ring and let P be a graded prime ideal of R. Then :

1. κ_{R-P} coincides on R-gr with κ_{C-p} where $p = P \cap C$ i.e. $\kappa_{R-P}(M) = \kappa_{C-p}({}_CM)$ for every $M \in R\text{-gr}$.
2. κ_{R-P} has property T and $Q^g_{R-P}(M) \cong Q^g_{C-p}(M)$ for all $M \in R\text{-gr}$.
3. If J is a graded ideal of R then $Q^g_{R-P}(R)\, j_\kappa(J)$ is a graded ideal of $Q^g_{R-P}(R)$ (here $j_\kappa : R \to Q^g_{R-P}(R)$ is the canonical graded morphism of degree o).
4. There is a one-to-one correspondence between graded prime ideals P of R such that $P \in \mathcal{L}(\kappa_{R-P})$, and proper graded prime ideals of $Q^g_{R-P}(R)$.

Consequently, if I is a graded ideal of R (hence rad(I) is a graded ideal of R too), then :

$$Q^g_{R-P}(R)\, j_\kappa(\mathrm{rad}(I)) = \mathrm{rad}(Q^g_{R-P}(R)\, j_\kappa(I)) \ .$$

Proof : 1. Since κ_{R-P} is induced on R-gr by $\underline{\kappa}_{R-P}$ and since $\underline{\kappa}_{R-P}$ and $\underline{\kappa}_{C-P}$ coincide on R-mod. (Theorem 15.1.2.)

it follows that κ_{R-P} and κ_{C-P} coincide on R-gr. Consequently κ_{R-P} may be regarded as being the kernel functor on R-gr associated to the central (Ore) set h(C-p).

2. From 1 it follows that, $I \in \mathcal{L}(\kappa_{R-P})$ if and only if there is an $x \in I \cap h(C-P)$, but then $x^{-1} \in Q^g_{R-P}(R)$ follows from $x^{-1} \in Q_{R-P}(R)$ and $(Rx)x^{-1} \subset R$ with $Rx \in \mathcal{L}(\kappa_{R-P})$. Thus $Q^g_{R-P}(R)\ j_{R-P}(I) = Q^g_{R-P}(R)$, where $j_{R-P} : R \to Q^g_{R-P}(R)$ is the canonical epimorphism of degree o. For $M \in$ R-gr, $Q_{R-P}(\underline{M}) \cong Q_{C-P}(\underline{M})$ follows from Zariski centrality of R. Since both $\underline{\kappa}_{R-P}$ and $\underline{\kappa}_{C-P}$ are graded and of finite type it follows that (cf. Proposition 12.4.7) $Q^g_{R-P}(M) \cong Q_{R-P}(\underline{M}) \cong Q_{C-P}(\underline{M}) \cong Q^g_{C-P}(M)$,

3. Let J be any graded ideal of R and take $x \in Q^g_{R-P}(R) j_{R-P}(J) Q^g_{R-P}(R)$. Write $x = \sum_i{}' q_i j_i q_i'$ with $q_i, q_i' \in h(Q^g_{R-P}(R))$, $j_i \in h(J)$. We may fix an element $c \in h(C-P)$ such that $cq_i' \in j_{R-P}(R)$ for every i. Since c is central in $Q^g_{R-P}(R)$, we get : $cx \in Q^g_{R-P}(R) j_{R-P}(J)$ or $x \in Q^g_{R-P}(R) j_{R-P}(J)$ by left multiplication with $c^{-1} \in Q^g_{R-P}(R)$.

4. For all graded prime ideals Q of R we have that $Q^e = Q^g_{R-P}(R) j_{R-P}(Q)$ is an ideal of $Q^g_{R-P}(R)$, which is non-proper if and only if $Q \in \mathcal{L}(R-P)$. It is easily checked that Q^e is a graded ideal and that $Q^e \cap j_{R-P}(R) = j_{R-P}(Q)$ if $Q \notin \mathcal{L}(R-P)$. If I is an ideal of $Q^g_{R-P}(R)$ which is not in Q^e then $I \cap j_{R-P}(R)$ is an ideal of $j_{R-P}(R)$ which is not in $j_{R-P}(Q)$ because $Q^g_{R-P}(R)(I \cap j_{R-P}(R)) = I$ by property T for κ_{R-P}.
Therefore, if I and J are ideals in $Q^g_{R-P}(R)$ such that $IJ \subset Q^e$ then $(I \cap j_{R-P}(R))(J \cap j_{R-P}(R)) \subset Q^e \cap j_{R-P}(R) = j_{R-P}(Q)$ and this entails that $I \cap j_{R-P}(R)$ or $J \cap j_{R-P}(R)$ is contained in $j_{R-P}(Q)$ since the latter is a prime ideal. Thus, I or J is contained in Q^e. Combination of the foregoing properties yields that every prime ideal of $Q^g_{R-P}(R)$ equals Q^e for some graded prime ideal Q of R. The second statement in 4. is a trivial consequence of the first. □

<u>15.2.2. Remarks</u>. 1. The same holds for ZG-rings.

2. In the same way similar properties may be established for any kernel functor associated to a central multiplicative set. In particular, such properties hold on a basis for the Zariski topology of Spec R, indeed one may consider κ_{Rc}, $c \neq o$ in C, which is associated to $\{1, c, c^2, \ldots\}$.

3. From the general theory of graded localizations it follows that : if R is prime

then $Q^g_\kappa(R)$ is prime for any rigid kernel functor on R-gr, if κ has property T then $Q^g_\kappa(R)$ is left Noetherian if R is left Noetherian.

15.2.3. Proposition. Let R be a positively graded Zariski central ring. Then R_0 is a Zariski central algebra.

Proof : Let I_0 be an arbitrary ideal of R_0. Clearly C_0 is in the centre of R_0. If P_0 is a prime ideal of R_0 such that $P_0 \supset R_0(I_0 \cap C_0)$ then $P_0 + R_+$ is a prime ideal of R such that :

$$P_0 + R_+ \supset \mathrm{rad}(R_0(I_0 \cap C_0) + R_+) = \mathrm{rad}(R((R_0(I_0 \cap C_0) + R_+) \cap C))$$

Now, $(R_0(I_0 \cap C_0)+R_+) \cap C \supset RI_0R \cap C$ is easily checked by compairing points of arbitrary degree; therefore

$$P_0 + R_+ \supset \mathrm{rad}(R(RI_0R \cap C)) = \mathrm{rad}(RI_0R) ,$$

hence $P_0 \supset I_0$. □

The first main advantage of Zariski central rings is that, even in the non left Noetherian case, Proj is having the nice properties one can hope for. First let us establish the following extended version of Theorem 14.2.

15.2.4. Theorem. Let R be a positively graded Zariski central ring. Let $n_0 \in \mathbb{N}$ be fixed, and suppose we are given additive subgroups p_n of R_n for all $n \geq n_0$. The following statements are equivalent :

1. There is a unique graded prime ideal P of R, $P \not\supset R_+$, such that $P \cap R_n = p_n$ for all $n \geq n_0$.

2.a. For all $n,k \in \mathbb{N}$ and for all $m \geq n_0$ we have $R_n p_m R_k \subset p_{n+m+k}$.

b. If $r \in R_n$, $t \in R_m$ with $n,m \geq n_0$ are such that $rR_k t \subset p_{n+m+k}$ for all $k \in \mathbb{N}$ then $r \in p_n$ or $t \in p_m$.

c. $p_n \neq R_n$ for some $n \geq n_0$.

Proof : Repeat the proof of Theorem 14.2. for 1 ⇒ 2. For 2 ⇒ 1, repeat the construction of the graded ideal P, $P \not\supset R_+$ as in Theorem 14.2. If we prove that P is a prime ideal then the uniqueness argument used in Theorem 14.2 stays valid and thus the proof will be finished. So, suppose that $s \in R_\lambda$, $w \in R_\mu$ are such that $sR_k w \subset P'_{\lambda+\mu+k}$ for all $k \in \mathbb{N}$. Put $T = RsR \cap C$, then $RTRw \subset P$. Suppose that $w \notin P$ yields $T \in P$ then it also yields $\mathrm{rad}(RT) \subset P$ i.e. $RsR \subset P$ or $s \in P$. This means that in the proof of Theorem 14.2 we may assume that $s \in C$ but then $RsRw$ is finitely generated (by sw) and the same argumentation carries over. □

The equivalent of Proposition 14.4 does hold for Zariski central but not necessarily left Noetherian rings, the proof goes by reduction to the commutative case.

II.16. SCHEME STRUCTURE OF Proj OVER A GRADED ZARISKI CENTRAL RING

16.1. Lemma. Let R be a positively graded Zariski central ring with centre C and let κ_C be a rigid kernel functor on C-gr. To κ_C we may associate a rigid kernel functor κ on R-gr, which is given by $\mathcal{L}(\kappa)$, the filter generated by the ideals RI, $I \in \mathcal{L}(\kappa_C)$. Then Q^g_κ is a central graded localization and κ has property T; moreover the centre of $Q^g_\kappa(R)$ is the graded ring of quotients of C at κ_C.

Proof : The first assertions are just modifications of the properties mentioned in Section II. 15.2. The statement about the centre is a straightforward consequence of the fact that $j^{-1}(Z)$, where Z is the centre of j(R) and where $j : R \to Q^g_{\kappa_C}(R)$ is the canonical ring homomorphism of degree o, is a graded C-module such that $j^{-1}(Z)/C$ is κ_C-torsion ; hence the localizations in R-gr of $j^{-1}(Z)$ and of C at κ_C, coincide. □

16.2. Lemma. Let R be a positively graded GZ-ring, then Proj R has a basis for the Zariski topology which consists of open sets $X_+(Rc)$ where c runs through the set of homogeneous elements of C_+. The graded kernel functor κ_{Rc} on R-gr, associated to $X_+(Rc)$, has property T. It is clear that κ_{Rc} is obtained from κ_{Cc} in the way described in Lemma 16.1. Let Q^g_c be the localization functor corresponding to κ_{Rc}, let $j_c : R \to Q^g_c(R)$ be the canonical ring morphism. Then graded ideals I of R extend to graded ideals of $Q^g_c(R)$. Furthermore $Q^g_c(R)$ is a Zariski central graded ring with centre $Q^g_{Cc}(C)$, Q^g_{Cc} being the localization functor on R-gr corresponding to κ_{Cc}.

As a consequence of Proposition 15.2.3. we have

16.3. Lemma. Let R be a positively graded GZ-ring. If p is a prime ideal of C_o, C being the centre of R, such that $C_o \cap R_o p = p$ then $\mathrm{rad}(R_o p)$ is the unique prime ideal of R_o lying over p.

16.4. Lemma. Let R be a positively graded Zariski central ring with centre C. Let κ be a rigid kernel functor on C-gr associated to a multiplicative set in C and let Q^g_κ be the localization functor on C-gr corresponding to κ. Denote $S = Q^g_\kappa(R)$, $D = Q^g_\kappa(C)$. Suppose that q is a graded prime ideal of C and put $q^{(e)} = \bigoplus_m q_{me}$, for some fixed $e \in \mathbb{N}$. Then we have $\mathrm{rad}(Sq) = \mathrm{rad}(Sq^{(e)})$.

Proof : The inclusion $\mathrm{rad}(Sq) \supset \mathrm{rad}(Sq)^{(e)}$ is clear. Conversely, since q is central, it follows that for any $x \in q_m$ we have that $x^e \in q_{me} \in q^{(e)}$, whence $x \in \mathrm{rad}(Cq^{(e)})$, thus $q \subset \mathrm{rad}(Cq^{(e)})$ follows. For any ideal I of R we have that : $Q^g_\kappa(R)\ \mathrm{rad}(I) = \mathrm{rad}(Q^g_\kappa(R) I)$ (cf. Theorem 15.2.1, 4.). Applying this to $\mathrm{rad}(Rq^{(e)})$ yields $Sq \subset \mathrm{rad}(Sq^{(e)})$ and thus $\mathrm{rad}(Sq) = \mathrm{rad}(Sq^{(e)})$ holds. □

The last lemma we need is the non-Noetherian version of the Stalk Theorem 14.1

16.5. Lemma. Let R be a positively graded Zariski central ring. Let Proj R be the topological space with its structural sheaf as constructed in Section 14. Then the stalk of Proj at $P \in \mathrm{Proj}\ R$ is exactly $Q^g_{R-P}(R)$.

Proof. There is a basis for the topology consisting of open sets $X_+(I)$ such that the graded rigid kernel functor K_I associated to $X_+(I)$ is central and $Q^g_I(R) \cong Q^g_{I \cap C}(R)$. (ring isomorphism of degree o). Consequently $\varinjlim_{P \in X_+(I)} Q^g_I(R)$ equals $\varinjlim_{P \in X_+(I) \in \beta} Q^g_I(R)$, where β is the selected basis for the topology, hence

$$\varinjlim_{P \in X_+(I)} Q^g_I(R) = \varinjlim_{p \in Y(I \cap C)} Q^g_{I \cap C}(R) = Q^g_{C-p}(R) \quad \text{(where } Y = \mathrm{Proj}\ C)$$

Since R is graded Zariski central $Q^g_{C-p}(R) \cong Q^g_{R-P}(R)$ and the statement follows. □

16.6. Theorem. 1. If R is a prime positively graded Zariski central ring then Proj R is a scheme. This means that there exists a basis β for the Zariski topology of Proj R, consisting of open sets $X_+(I)$ where I is a graded ideal of R contained in R_+, such that each $X_+(I)$ endowed with the induced topology and sheaf is isomorphic to $\mathrm{Spec}(Q_I^g(R))_0$ with its usual topology and structural sheaf (cf. [32]). If $X_+(I)$ and $X_+(J)$ are in β then the ring $(Q_{IJ}^g(R))_0$ is generated as a ring by the restrictions of the rings $(Q_I^g(R))_0$ and $(Q_J^g(R))_0$.

2. If R is a positively graded Zariski central ring, then the presheaves of $X_+(I)$ and $\mathrm{Spec}(Q_I^g(R))_0$ are isomorphic. In this case the scheme structure of Proj R follows by sheafification methods.

Proof : For the basis β we will choose $\{X_+(Rc)$, c a homogeneous element of $C_+\}$. Since R is a Zariski central ring this β is a basis for the topology on Proj R as well as a covering for it. We split the proof in the following three parts :

a. The bijection between $X_+(Rc)$ and $\mathrm{Spec}(Q_c^g(R))_0$,

b. The topological homeomorphism, and

c. The (pre-) sheaf isomorphism.

a. Bijective correspondence between $X_+(Rc)$ and $\mathrm{Spec}(Q_c^g(R))_0$.

Let $q_0 \in \mathrm{Spec}(Q_c^g(R))_0$. Since $Q_c^g(R)$ is $Q_c^g(C)$-Zariski central, it follows from Proposition 15.2.3. that $(Q_c^g(R))_0$ is Zariski central over $(Q_c^g(C))_0$ (which is contained in, but not necessarily equal to the centre of $(Q_c^g(R))_0$. So if $p_0 = q_0 \cap (Q_c^g(R))_0$ then p_0 is prime and q_0 is the unique prime ideal of $(Q_c^g(R))_0$ lying over p_0 i.e. : $q_0 = \mathrm{rad}((Q_c^g(R))_0 p_0)$.

Define $q' \subset C$ by, $q'_m = \{d \in C_m,\ d^e c^{-m} \in q_0,\ e = \deg(c)\}$. It is easily verified that q' remains unaltered if one substitutes p_0 for q_0 in this definition. Now let us first establish that $q'_m = \{d \in C_m$, there exist $N, M \in \mathbb{N}$ with $eM = mN$ and $d^N c^{-M} \in q_0\}$. That q'_m is contained in the latter set is obvious. Conversely, suppose $N \geq e$, then $M \geq m$ and $(d^e c^{-m}).(d^{N-e} c^{-M+m}) \in q_0$. Since q_0 is prime and both elements considered are central it follows that either $d^e c^{-m} \in q_0$ or else $d^{N-e} c^{-M+m} \in q_0$. In the first case we are done. In the second case we repeat the proces and in the end we have to face the case $N < e$, $M < m$ with $eM = Nm$. Then : $(d^{e-N} d^N)(c^{M-m} c^{-M}) = (d^{e-N} c^{M-m}).(d^N c^{-M})$ is an element of

$(Q_C^g(C))_o\, q_o \subset q_o$, thus also $d^e c^{-m} \in q_o$. The characterization of q' just obtained makes it clear that q' is a graded prime ideal of C.

Consider $C \cap Rq'$ and pick $c_1 = \sum_i r_i y_i \in C \cap Rq'$, with $y_i \in q'$, $r_i \in R$ and, as $C \cap Rq'$ is graded, we may suppose that c_1, r_i, y_i are homogeneous elements. Since y_i commutes with r_j for all (i,j), we may choose N large enough such that $c_1^N = \Sigma\, r_{i_1} \dots r_{i_N}\, y_1^{\nu_1} \dots y_i^{\nu_i - e}\, y_i^e$ with $\nu_i > e$, i.e. each term in the sum has the form given above with respect to at least one index i appearing in the expression for c_1. Enlarging N if necessary, we may assume that $N = e.\nu$.

Thus $c_1^N c^{-\nu m}$ with $m = \deg(c_1)$ is in $(Q_C^g(R))_o\, q_o \subset q_o$ because each term in $c_1^N c^{-\nu m}$ may be written in the form :

$$c^{-\nu m + \deg(y_i)} \cdot r_{i_1} \dots r_{i_N}\, y_1^{\nu_1} \dots y_i^{\nu_i - e} \cdot y_i^e\, c^{-\deg y_i}$$

which obviously is in q_o because $y_i \in q$.

It follows that $Rq' \cap C = q'$, hence $\mathrm{rad}(Rq') = Q$ is a prime ideal of R by Zariski centrality of R. Notation : write Q^{ex} for $Q_C^g(R)Q$.

By Theorem 15.2.1. : $Q^{ex} = \mathrm{rad}(Q_C^g(R)q')$ and then by Lemma 16.4 : $Q^{ex} = \mathrm{rad}(Q_C^g(R)q'^{(e)})$. Clearly Q is a graded prime ideal of R which does not contain c because $c \in Q$ yields $c \in q'$ and $1 \in q_o$. Thus $Q \in X_+(Rc)$ and we have obtained a well defined map $\psi : \mathrm{Spec}(Q_C^g(R))_o \to X_+(Rc)$, given by $q_o \to Q$. Conversely, given $Q \in X_+(Rc)$, consider $(Q^{ex})_o \cap (Q_C^g(C))_o$. If y is in the latter set then $c^n y \in Q \cap C_{ne}$ for some n. Put $q' = Q \cap C$ and form $q'^{(e)}$. Then $p_o = \{d \in (Q_C^g(C))_o,\ c^n d \in q'_{ne}$ for some n$\}$, is a prime ideal of $(Q_C^g(C))_o$. So the relation $c^n y \in Q \subset C_{nc}$ translates to $y \in p_o$ i.e. $(Q^{ex})_o \cap (Q_C^g(C))_o = p_o$. By Zariski centrallity of $(Q_C^g(R))_o$ over $(Q_C^g(C))_o$ it follows then that $\tilde{q}_o = \mathrm{rad}((Q^{ex})_o)$ is a prime ideal of $(Q_C^g(R))_o$. One checks that Q corresponds to $\tilde{q}_o$ in the way described first, hence the map ψ is clearly surjective. Injectivity of ψ will follow from, $q_o = \tilde{q}_o = \mathrm{rad}((Q^{ex})_o)$ where $\psi(q_o) = Q$ and $\psi(\tilde{q}_o) = Q$. It has been established already that $Q^{ex} = \mathrm{rad}(Q_C^g(R)q'^{(e)})$. Now take $x \in (Q^{ex})_o$. Since in any positively graded ring T any ideal I has the property $\mathrm{rad}(I_o) \supset (\mathrm{rad}\ I)_o$ we get that $x \in \mathrm{rad}((Q_C^g(R)q'^{(e)})_o)$.

However we have :

$$(Q_c^g(R)q'^{(e)})_o = (Q_c^g(R))_o\, q_o' + \sum_i (Q_c^g(R))_{-ie}\, q_{ie}' \quad .$$

But $y \in (Q_c^g(R))_{-ie}$ means that $c^i y \in (Q_c^g(R))_o$, hence $(Q_c^g(R))_{-ie} = (Q_c^g(R))_o\, c^{-i}$, whereas for each i, $z \in q'_{ie}$ means that $zc^{-i} \in p_o$. Taking into account that $q_o' \subset p_o$ we obtain $x \in \mathrm{rad}((Q_c^g(R))_o\, p_o) = q_o$. Finally, we arrive at : $p_o \subset (Q^{ex})_o \subset q_o$ and therefore $q_o = \mathrm{rad}((Q^{ex})_o)$ follows from the fact that q_o is prime and $p_o = q_o \cap (Q_c^g(C))_o$.

b. The homeomorphism $X_+(Rc)$ $\mathrm{Spec}(Q_c^g(R))_o$.

Since $X_+(Rcd) = X_+(Rc) \cap X_+(Rd)$, $c,d \in C_+$, the open sets $X_+(Rcd)$ with d varying through C_+ forms a basis for the topology induced in $X_+(Rc)$ by the Zariski topology of Proj R. Let $\psi : X_+(Rc) \to \mathrm{Spec}(Q_c^+(R))_o$, $P \to \mathrm{rad}((P^{ex})_o)$ be the bijective map constructed in a. Let $d \in C_m$, then $d^e c^{-m} \in (Q_c^g(R))_o$. If $P \in X_+(cd)$ then $P \in \mathrm{Proj}\ R$ is such that $id \notin P$. This implies that $d^e c^{-m} \notin \mathrm{rad}((P^{ex})_o)$; indeed $(d^e c^{-m})^N \in (P^{ex})_o$ would yield $d^{eN} \in P^{cx}$ or $c^M\, d^{eN} \in P$ for some $M \in \mathbb{N}$. Both c,d are central, thus $c \notin P$ yields $d \in P$ and $cd \in P$. Conversely, if $\psi(P)$ is in the open set $\{q \in \mathrm{Spec}(Q_c^g(R))_o$, q does not contain $d^e c^{-m}\}$, then it is equally easy to derive from this that $cd \notin P$. So it becomes evident that ψ is a topological homeomorphism, as claimed.

c. The sheaf isomorphism.

First consider the case where R is a prime ring. In this case the presheaf induced on $X_+(Rc)$ is actually a sheaf. On the other hand $(Q_c^g(R))_o$ need not be a prime ring, however since $(Q_c^g(R))_o$ is a Zariski central ring it is easy to prove that the usual presheaf on $\mathrm{Spec}(Q_c^g(R))_o$ is a separated presheaf. To prove the sheaf isomorphism in this case it will therefore be sufficient to find a basis for the topology such that over open sets in this basis, the ring of sections of the sheaf on $X_+(Rc)$ is isomorphic to the ring sitting over the image of that open set in the presheaf on $\mathrm{Spec}(Q_c^g(R))_o$. Let $c,d \in C_+$ and put $\deg(c) = e$, $\deg(d) = m$, $c' = c^m d^{-e} \in (Q_d^+(c))_o$ and $d' = d^e c^{-m} \in (Q_c^g(C))_o$. To the open set $X_+(Rcd)$ in $X_+(Rc)$ there corresponds the open set $Y(d') = \{q \in \mathrm{Spec}(Q_c^g(R))_o,$

$d' \notin q\}$ in $Y = \mathrm{Spec}(Q_c^g(R))_0$. In the (pre) sheaf structure of $X_+(Rc)$ the ring of sections over $X_+(Rcd)$ is $(Q_{cd}^g(R))_0$, while the presheaf on Y associates $Q_{d'}((Q_c^g(R))_0)$ to $Y(d')$, where $Q_{d'}$ is the localization functor in $(Q_c^g(R))_0$-mod corresponding to the central multiplicatively closed set generated by d'. It will now be sufficient to prove the following :

$$Q_{d'}((Q_c^g(R))_0) \cong (Q_{cd}^g(R))_0 \cong Q_{c'}((Q_d^g(R))_0).$$

Since c and d commute, the second isomorphism will follow from the first by a symmetric argument.

Define a ring morphism $\varphi : Q_{d'}((Q_c^g(R))_0) \to (Q_{cd}^g(R))_0$ by

$\varphi[(d')^{-N}(xc^{-f})] = (x\, c^{e+m}\, d^f)(cd)^{-(f+eN)}$, where $x \in (Q_c^g(R))_{fe}$.

The inverse homomorphism $\psi : (Q_{cd}^g(R))_0 \to Q_{d'}((Q_c^g(R))_0)$ will then be given by

$\psi(y.(cd)^{-f}) = (d')^{-M}.(y\, d^{f(e-1)})c^{-f(m+1)}$, where $y \in (Q_{cd}^g(R))_{f(e+m)}$. All implicit assumptions may easily be verified (this part of the proof is identical to the proof in the commutative case).

That $(Q_{cd}^g(R))_0$ is generated by the images of $(Q_c^g(R))_0$ and $(Q_d^g(R))_0$ is again mere verification. In case R is not prime, one may repeat a similar proof, taking care to replace R by $j_c(R)$ or $j_d(R)$ where necessary. Since torsion parts will behave nice we will obtain, over a basis of the topology, isomorphisms between the rings making up the presheaves on $X_+(Rc)$ and $\mathrm{Spec}(Q_c^g(R))_0$. Sheafification then yields the desired result. □

<u>16.7. Remark.</u> If in the situation of the theorem we have that $R_0 \subset C$ i.e. $R_0 = C_0$ then it follows that $R_0 \to (Q_c^g(R))_0$ is an extension of R_0 and therefore, to the canonical ring morphism $R_0 \to (Q_c^g(R))_0$ there corresponds a presheaf morphism : $\mathrm{Spec}(Q_c^g(R))_0 \to \mathrm{Spec}\ R_0$. In this case Proj R may be viewed as a Spec R_0-scheme.

<u>16.8. Remark.</u> If in the situation of the theorem C is generated as an C_0-ring by C_1 then we have :

1. If $P \in X_+(Rc)$ then $(P^{ex})_0$ is a prime ideal of $(Q_c^g(R))_0$ i.e. the bijection ψ may then be given by $\psi(P) = (P^{ex})_0$.

2. If R is a prime ring then $(Q_C^g(R))_o$ is prime and the presheaf $\mathrm{Spec}(Q_C^g(R))_o$ is a sheaf.

Proof : 2. will follow from 1. applied to $P = o$.

1. Pick $c \in C_+$. It has been noted that $X_+(Rc)$ depends only on $\mathrm{rad}((Rc)_+)$, hence the fact that $C_m = C_1^m$ implies that we get a basis for the Zariski topology of Proj R by taking $\{X_+(Rc), c \in C_1\}$. If I and J are ideals of $(Q_C^g(R))_o$ such that $IJ \subset (P^{ex})_o$ for some $P \in X_+(Rc)$ then pick $i \in I$, $j \in J$. For any $q \in (Q_C^g(R))_m$ we see that $c^{-m} q \in (Q_C^g(R))_o$ hence $iqj \in c^m (P^{ex})_o \subset (P^{ex})_m$. Since P^{ex} is a graded prime ideal of $Q_C^g(R)$ it follows that i or j is in P^{ex}; hence i or j is in $(P^{ex})_o$.

16.9. Remark. The statement of 16.8 holds for Azumaya algebras over polynomial rings but fails for rings of twisted polynomials unless these are already Azumaya algebras. Indeed if $R = A[X,\sigma]$, where A is a simple ring, has center $k[T]$ then $T = X$ implies that R is an Azumaya algebra thus if not, then T is a polynomial in X of degree higher than 1.

III. LOCAL CONDITIONS FOR NOETHERIAN GRADED RINGS.

In this chapter, all rings considered will be commutative and Noetherian unless explicitely mentioned otherwise.

III.1. Injective Dimension of Graded Rings.

Let us introduce the following terminology : a ring R is termed a gr.-local ring if it has exactly one maximal graded ideal (these ideals are said to be gr.-maximal) in the set of proper graded ideals of R.

Let P be a prime ideal of R and let S be the multiplicatively closed set $h(R) - P = h(R-P)$. Since R is Noetherian, our results in II.12 and II.13 imply that localization in R-mod at the kernel functor associated to S is inner in R-gr i.e. for any $M \in$ R-gr , $Q_S(\underline{M}) = \underline{Q^g_{R-P}(M)}$. If $Q = (P)_g$ then $S = h(R)-Q$ and $Q_S(\underline{M}) = \underline{Q^g_{R-Q}(M)}$ follow. Since $\underline{\kappa}_S$ has property T in R-mod, κ_S has property T in R-gr and therefore $Q^g_{R-P}(R)$ is gr.-local and $Q^g_{R-P}((P)_g) = Q^g_{R-P}(R)(P)_g$ is gr-maximal.

Let $k^g(P)$ denote the graded division ring $Q^g_{R-P}(R)/Q^g_{R-P}(P)$.

1.1. Lemma. Let P be a prime ideal of R, then :

1. For any $M \in R$-gr, $Q_{R-P}(\underline{M}) = o$ if and only if $Q_{R-(P)_g}(\underline{M}) = o$ if and only if $Q^g_S(M) = o$.
2. Put supp $M = \{P \in \text{Spec } R,\ Q_{R-P}(\underline{M}) = o\}$, then $P \in$ supp M if and only if $(P)_g \in$ supp M.
3. If P is a graded prime ideal then : $E^g_{Q^g_{R-P}(R)}(k^g(P)) \cong E^g_R(R/P)$.
4. Let P be a graded prime ideal and let $A_i = \{x \in E^g(R/P),\ P^i x = o\}$ then the following properties hold :
   a. A_i is a graded R-submodule of $E^g(R/P)$ and $E^g(R/P) = \bigcup_{i \geq 1} A_i$.
   b. A_{i+1}/A_i is a graded $k^g(P)$-module.
   c. $A_1 \cong k^g(P)$.
5. Any injective object in R-gr is a direct sum of some $E^g(R/P)(n)$, for some graded prime ideals P of R and some $n \in \mathbb{Z}$.
6. Denote by gr.inj.dim$_R M$ the injective dimension of M in the category R-gr. If $S \subset h(R)$ is any multiplicatively closed set, $M \in R$-gr, then : gr.inj.dim$_{S^{-1}R}\, S^{-1}M \leq$ gr.inj.dim$_R M$.

Proof : 1., 2., 3. are easy consequences of the foregoing chapters. 4., 5., 6., follows as in the ungraded case, cf. [22]. □

Let $M \in R$-gr. and consider minimal injective resolutions in R-gr and R-mod resp. :

$$o \to M \to E^g_o \to E^g_1 \to \cdots$$

$$o \to M \to E_o \to E_1 \to \cdots .$$

For a prime ideal P of R, let $\mu^g_n(P,M)$, resp. $\mu_n(P,M)$, be the number of copies of $E^g_R(R/P)$ in E^g_n, resp. of $E_R(R/P)$ in E_n.

1.2. Proposition. Let $M \in R$-gr, P a graded prime ideal of R. For notational convenience we write Q^g for the localization functor on R-gr associated to h(R)-P, then :

1. The group $Q^g\ \text{Ext}^n_R(R/P,M)$ is a free graded $k^g(P)$-module of rank $\mu^g_n(P,M)$.

2. $\mu_i^g(P,M) = \mu_i(P,M)$

3. $\text{inj.dim}_{Q^g(R)} Q^g(M) = \text{gr.inj.dim}_{Q^g(R)} Q^g(M)$.

Proof : We have

$$\text{Ext}_R^i(R/P,M) = \text{EXT}_R^i(R/P,M) = \text{HOM}_R^i(R/P,Q_i^g) = \text{Hom}_R(R/P,Q_i^g) ,$$

hence :

$$Q^g \text{ Ext}_R^i(R/P,M) = Q^g\text{Hom}_R(R/P,Q_i^g) = \text{Hom}_{Q^g(R)}(k^g(P),Q_i^g)$$

and the latter is a free $k^g(P)$-module.

If P_1 is a graded prime ideal of R such that $P_1 \neq P$ then $Q^g \text{ Hom}(R/P, E_R^g(P_1)) = 0$, hence $Q^g \text{ Hom}_R(R/P,Q_i^g)$ is a free $k^g(P)$-module of rank equal to $\mu_i^g(P,M)$. Since $\text{Ext}_R^i(R/P,M)$ is a free $k(P)$-module of rank $\mu_i(P,M)$ whilst $Q \text{ Ext}(R/P,M)$ is the localization of $Q^g \text{ Ext}(R/P,M)$ at P (Q the localization in R-mod associated to P), 2. follows immediatly and 3 follows from 2. □

1.3. Proposition. Let P be a prime ideal of R such that $P \neq (P)_g$, then for all $n \in \mathbb{N}$ and $M \in R\text{-gr}$ we obtain : $\mu_n((P)_g,M) = \mu_{n+1}(P,M)$.

Proof : Because of Corollary 2.4 in [1] we have : $\mu_n((P)_g,M) = \mu_n(Q^g((P)_g),Q^g(M))$ while $\mu_{n+1}(P,M) = \mu_{n+1}(Q^g(P),Q^g(M))$. So we may replace R and M by $Q^g(R)$ and $Q^g(M)$ resp., i.e. we may suppose that R is gr-local and P the gr.-maximal ideal of R. In that case P is a maximal ideal of R. Since $R/(P)_g$ is a graded division ring, there exists $a \in R$ such that $P = (P)_g + Ra$. From the exact sequence :

$$(\star) \qquad 0 \to R/(P)_g \xrightarrow{m_a} R/(P)_g \to R/P \to 0 ,$$

where m_a denotes multiplication by a, we obtain :

$$\ldots \to \text{Ext}^n(R/(P)_g,M) \xrightarrow{m_a} \text{Ext}^n(R/(P)_g,M) \to \text{Ext}^{n+1}(R/P,M) \to$$

$$\ldots \to \text{Ext}^{n+1}(R/(P)_g,M) \xrightarrow{m_a} \text{Ext}^{n+1}(R/(P)_g,M) \to \ldots$$

Note that $\text{Ext}^n(R/(P)_g,M) = \text{EXT}^n(R/(P)_g,M)$ is a graded $R/(P)_g$-module, hence $\text{Ext}^n(R/(P)_g,M)$

is a free graded $R/(P)_g$ module. Consequently,

$$m_a : \mathrm{Ext}^n(R/(P)_g,M) \to \mathrm{Ext}^n(R/(P)_g,M) \ ,$$

is monomorphic. The long exact sequence obtained above yields a short exact sequence :

$$o \to \mathrm{Ext}^n(R/(P)_g,M) \xrightarrow{m_a} \mathrm{Ext}^n(R/(P)_g,M) \to \mathrm{Ext}^{n+1}(R/P,M) \to o \ .$$

In this sequence $V = \mathrm{Ext}^n(R/(P)_g,M)$ is free graded of rank $\mu_n((P)_g,M)$, whereas $V' = \mathrm{Ext}^{n+1}(R/P,M)$ is a vector space of dimension $\mu_{n+1}(P,M)$ over the field R/P. In $R/(P)_g$-mod we have the exact sequence :

$$(\star\star) \qquad o \longrightarrow V \xrightarrow{m_a} V \longrightarrow V' \longrightarrow o \ .$$

Since $V \cong (R/(P)_g)^{\mu_n((P)_g,M)}$ we deduce from $(\star)$ and $(\star\star)$ that $V' \cong (R/P)^{\mu_n((P)_g,M)}$. On the other hand $V' = (R/P)^{\mu_{n+1}(P,M)}$, therefore $\mu_n((P)_g,M) = \mu_{n+1}(P,M)$. □

<u>1.4. Corollary</u>. In the situation of 1.3, $\mu_n((P)_g,M) = o$ if and only if $\mu_{n+1}(P,M) = o$.

<u>1.5. Corollary</u>. Let P be a graded prime ideal of R. The minimal injective resolution of $E^g(R/P)$ in R-mod is the following :

$$o \longrightarrow E^g(R/P) \longrightarrow E(R/P) \longrightarrow \bigoplus_{P'} E(R/P') \longrightarrow o$$

where the direct sum is over all prime ideals $P' \neq P$ of R such that $(P')_g = P$.

<u>Proof</u> : That resolution has the form :

$$o \longrightarrow E^g(R/P) \longrightarrow E(R/P) \longrightarrow Q_1 \longrightarrow o$$

where $Q_1 = \sum_{P' \in \mathrm{Spec}\, R} E(R/P')^{\mu_1(P',E^g(R/P))}$. Applying Proposition 1.3. and Corollary 1.4. we get $Q_1 = \bigoplus_{P'} E(R/P')$ where $(P')_g = P$. □

<u>1.6. Remark</u>. If M is a maximal ideal of R which is graded then $E^g_R(R/M) = E_R(R/M)$, e.g. if $R = k[X_1 \dots X_n]$ and $M = (X_1 \dots X_n)$.

1.7. Corollary. Let R be a commutative Noetherian ring with left limited grading. If $M \in R$-gr is a finitely generated graded module then : $\text{gr.inj.dim}_R M = \text{inj.dim}_{\underline{R}} \underline{M}$.

Proof : Put $n = \text{gr.inj.dim}_R M < \infty$. If $\text{inj.dim}_{\underline{R}} \underline{M} \neq n$ then $\text{inj.dim}_{\underline{R}} \underline{M} = n+1$. Now $\mu_{n+1}(P,M) \neq o$ for some prime ideal P of R, so by the foregoing results P is maximal. Clearly P is not graded and by Proposition 1.3. it results that $\mu_n((P)_g,M) \neq o$ hence $\mu^g_n((P)_g,M) \neq o$. We claim that $(P)_g$ is graded maximal, indeed if not, then $(P)_g \subset P_1$, P_1 graded yields that $\mu^g_{n+1}(P_1,M) \neq o$, contradiction. Since R has left limited gradation it follows that $(P)_g$ is a maximal ideal, i.e. $P = (P)_g$, contradiction. Consequently, $\text{inj.dim}_{\underline{R}} \underline{M} = n$.

In going from R to R[X] we need the following lemma where R has trivial gradation and R[X] is graded as usual :

1.8. Lemma. If P is a graded prime ideal of R[X] then either $P = pR[X]$ or $P = pR[X] + (X)$, where $p = P \cap R$.

Proof : Since $P \supset pR[X]$ and $R[X]/pR[X] = (R/P)[X]$ we may assume $p = o$, i.e. R is a domain. Let K be the field of fractions of R. Since $P \cap R = o$, $PK[X]$ is a graded prime ideal of $K[X]$ i.e. $PK[X] = o$ or $PK[X] = (X)$. □

1.9. Proposition. Let R be a Noetherian commutative ring and take $P \in \text{Spec } R$. Put $E = E_R(R/P)$. Consider the ring R[X] graded corresponding to the degree in X. Denote $M = E[X] = E \otimes_R R[X]$.
The minimal injective resolution of M in R[X]-gr is :

$$o \to M \to E^g_R(R[X]/PR[X]) \to E^g(R[X]/PR[X] + (X)) \to o$$

and the minimal injective resolution of M in R[X]-mod is :

$$o \to \underline{M} \to E_{R[X]}(R[X]/PR[X]) \to \bigoplus_{P'} E_{R[X]}(R[X]/P') \to o$$

the direct sum being taken over the prime ideals P' of R[X] such that $P' \cap R = P$.

Proof : Consider the minimal injective resolution of M in R[X]-gr :

$$o \longrightarrow M \longrightarrow I_0^g \longrightarrow I_1^g \longrightarrow I_2^g \longrightarrow \dots$$

Since Ass $M = \{PR[X]\}$, $I_0^g = E^g(R[X]/PR[X])$. Taking into account, Corollary 2.6. [1], we obtain :

$$\mu_{i+1}(PR[X] + (X), M) = \mu_i(PR[X] + (X)/(X),\ E[X]/XE(X)) = \mu_i(P,E)\ .$$

Since $\mu_i(P,E) = 1$ if $i = o$ and $\mu_i(P,E) = o$ if $i > o$, we have $\mu_1(PR[X] + (X), M) = 1$ and $\mu_i(PR[X] + (X), M) = o$ for all $i > 1$. Let P' be a prime ideal of $R[X]$ such that $\mu_i(P', M) > o$. Then necessarily $P' \in \mathrm{Supp}\ M$ and therefore $P' \cap R \supset P$. If $P' \cap R \neq P$, put $P' \cap R = Q$. By Lemma 1.8. it follows that $P' = QR[X]$ or $P' = QR[X] + (X)$.
In case $P' = QR[X]$ we have :

$$\mu_i(P', M) = \mu_i(QR[X], E[X]) = \dim_{k(P)} Q_{R-P}(\mathrm{Ext}^i_{R[X]}(R[X]/QR[X]\ ,\ E[X]))\ .$$

However,

$$\mathrm{Ext}^i_{R[X]}(R[X]/QR[X], E[X]) = \mathrm{Ext}^i_R(R/Q,E) \underset{R}{\otimes} R[X]$$

and as $Q \neq P$ we obtain $\mathrm{Ext}^i_R(R/Q,E) = o$ for all $i \geqslant o$. Thus $\mu_i(P', M) = o$ for all $i \geqslant o$.
In case $P' = QR[X] + (X)$, we have an exact sequence

$$o \to R[X]/QR[X] \xrightarrow{m_X} R[X]/QR[X] \to R[X]/P' \longrightarrow o$$

yielding a long exact sequence :

$$\dots\ \mathrm{Ext}^i(R[X]/QR[X], M) \xrightarrow{m_X} \mathrm{Ext}^i(R[X]/QR[X], M) \to \mathrm{Ext}^{i+1}(R[X]/P', M) \to$$

$$\to \mathrm{Ext}^{i+1}(R[X]/QR[X], M) \xrightarrow{m_X} \mathrm{Ext}^{i+1}(R[X]/GR[X], M) \to$$

(all Ext are $\mathrm{Ext}_{R[X]}$).
From the above we deduce : $\mathrm{Ext}^i(R[X]/P', M) = o$ for all $i \geqslant 1$. Finally this amounts to $I_1^g = E^g(R[X]/PR[X] + (X))$ and $I_k^g = o$ for all $k > 1$.
For the second statement of the proposition, let the minimal injective resolution for $\underline{M}$ in $R[X]$-mod be given by :

$$o \longrightarrow \underline{M} \longrightarrow I_0 \longrightarrow I_1 \longrightarrow I_2 \longrightarrow \dots\ .$$

It is clear that $I_o = E_{R[X]}(R[X]/PR[X])$. From Proposition 1.2 and 1.3 it follows that I_1 contains $\oplus_{P'} E_{R[X]}(R[X]/P')$, where P' varies in the set of prime ideals of $R[X]$ such that $(P')_g = PR[X]$ or $P' = PR[X] + (X)$. Therefore $(P')_g = PR[X]$ if and only if $P' \cap R = P$. Indeed if $P' \cap R = P$ and $P' \neq (P')_g$ then $ht(P') = 1 + ht(P')_g)$, thus if $(P')_g \neq PR[X]$ then $ht(P')_g \geq 1 + ht(PR[X]) = 1 + ht(P)$. Thus $ht(P') \geq 2 + ht(P)$. On the other hand, since R is Noetherian and $P' \cap R = P$, $ht(P') \leq 1 + ht(P)$, cf. [17], contradiction. The converse implication being obvious this proves our statement.

Let P_1 be a prime ideal of $R[X]$ which is not graded and such that $P_1 \cap R \supsetneq P$. Then $(P_1)_g \supsetneq PR[X]$ and $(P_1)_g \neq PR[X] + (X)$ because $(P_1)_g = PR[X] + (X)$ would imply $P_1 \cap R = (P_1)_g \cap R = P$. From Proposition 1.3. we retain that $\mu_i((P_1)_g, M) = \mu_{i+1}(P_1, M))$. However we have already established $\mu_i((P_1)_g, M) = o$ for all $i \geq o$, hence $\mu_i(P_1, M) = o$ for all $i \geq 1$. This yields thus that $I_1 = \oplus_{P'} E_{R[X]}(R[X]/P')$ where P' is as stated. If I_2 were nonzero, then there exists a prime ideal P_2 of $R[X]$ such that $\mu_2(P_2, M) \neq o$. If P_2 is graded then from Proposition 1.2, we deduce $\mu_2^g(P_2, M) \neq o$ and thus $I_2^g \neq o$, contradiction.

If P_2 is not graded, then we deduce from Proposition 1.3. that $\mu_2(P_2, M) = \mu_1((P_2)_g, M) \neq o$. In this case it is necessary that $(P_2)_g = PR[X] + (X)$, hence $P_2 \cap R = (P_2)_g \cap R = P$. Since $P_2 \neq (P_2)_g$, $ht(P_2) = 1 + ht((P_2)_g) = 2 + ht(PR[X] = 2 + ht(P)$, but from $P_2 \cap R = P$, $ht(P_2) \leq 1 + ht(P)$ follows (cf. [17] as before), contradiction.

Therefore $\mu_2(P_2, M) = o$ and $I_2 = o$ follows.

<u>1.10. Corollary</u>. Let R be a commutative Noetherian ring, E injective in R-mod, then :

$$gr.inj.dim_{R[X]} E[X] + 1 = inj.dim_{R[X]} E[X] \quad .$$

III.2. Regular, Gorenstein and Cohen-Macaulay Rings.

Let R be a commutative Noetherian ring and M a nonzero finitely generated R-module. We put : $V(M) = supp(M) = V(Ann_R M) = \{P \in Spec\ R,\ Ann_R M \subset P\}$. Recall that the Krull dimension of M, denoted by $K.dim_R M$ is defined to be the supremum of the lengths of chains of prime ideals of $V(M)$ if this supremum exists, and ∞ if not. (the reader

may verify that $K.\dim_R M$ coincides with the Krull dimension of M defined in Section I.5.).
We have $ht_M P = K.\dim_{Q_{R-P}(R)}(Q_{R-P}(M))$.
If I is an ideal of R such that $IM \neq M$ then the least r for which $\mathrm{Ext}^r_R(R/I,M) \neq 0$ will be called the grade of I on M, denoted by : grade (I,M). It is easily checked that grade(I,M) is exactly the common length of all maximal sequences contained in I.

If R is a local ring with maximal ideal Ω, then M is said to be a Cohen-Macaulay module or a C.M.-module if grade $(\Omega,M) = K.\dim_R M$. A not necessarily local ring is said to be a Cohen-Macaulay ring or C.M.-ring if $Q_{R-P}(R)$ is a Cohen-Macaulay $Q_{R-P}(R)$-module and in that case M is a C.M.-module if $Q_{R.P}(M)$ is a C.M.-$Q_{R-P}(R)$-module. It is easy to verify that M is a C.M.-module if and only if for each maximal ideal $\Omega \in V(M)$, $\mu_i(\Omega,M) = 0$ whenever $i < ht_M(\Omega)$, cf. [37]. We say that M is a Gorenstein-module if for every maximal ideal $\Omega \in V(M)$, $\mu_i(\Omega,M) = 0$ if and only if $i \neq ht_M(\Omega)$. In [37] it has been established that M is a Gorenstein-module if and only if $Q_{R-P}(M)$ is a Gorenstein-$Q_{R-P}(R)$-module for all $P \in V(M)$.
A Gorenstein ring R is a ring which is a Gorenstein-module when considered as a module over itself. For a detailed study of these classes of rings we refer to [37]. First we give local-global theorems in the graded case.

2.1. Theorem. Let R be a commutative Noetherian graded ring, $M \in R$-gr a finitely generated object. The following statements are equivalent :

1. M is a Cohen-Macaulay module.
2. For any graded prime ideal P of V(M), $Q_{R-P}(M)$ is a Cohen-Macaulay $Q_{R-P}(R)$-module.

Proof : 1 ⇒ 2. is trivial. To prove the converse let us assume first that $Q_{R-P}(M)$ is a C.M. $Q_{R-P}(R)$-module for every graded $P \in V(M)$. Let $P' \in V(M)$. If $(P')_g = P'$ then we are done. So suppose that $P' \neq (P')_g$. In this case we have that $ht_M P' = ht_M (P')_g + 1$. Choose $i < ht_M(P')$. Then $i-1 < ht_M((P')_g)$, hence $\mu_{i-1}((P')_g,M) = 0$. By Proposition 1.3 it follows then that $\mu_i(P',M) = 0$ and therefore M is a C.M.-module. □

2.2. Theorem. Let R be a graded commutative Noetherian ring, $M \in R$-gr a finitely generate object. The following statements are equivalent :

1. M is a Gorenstein-module.
2. For each graded $P \in V(M)$, $Q_{R-P}(M)$ is a Gorenstein $Q_{R-P}(R)$-module.

Proof : 1 ⇒ 2. Easy. For the converse, suppose that $Q_{R-P}(M)$ is a Gorenstein $Q_{R-P}(R)$-module for each graded $P \in V(M)$. Let P' be arbitrary in $V(M)$. If $P' = (P')g$ then the assertion is clear. Supposing that $P' \neq (P')_g$ and $i \neq ht_M(P')$, we obtain $i-1 \neq ht_M((P')_g)$. Since $Q_{R-(P')_g}(M)$ is a Gorenstein $Q_{R-(P')_g}(R)$-module it follows that :

$$\mu_{i-1}((P')_g, M) = \mu_{i-1}(Q_{R-(P')_g}((P')_g), Q_{R-(P')_g}(M)) = o \quad .$$

From Proposition 1.3. we deduce that $\mu_i(P', M) = o$. In a similar way it may be deduced from $\mu_i(P', M) = o$ that $i \neq ht_M(P')$; hence M is a Gorenstein module. □

A local ring with maximal ideal Ω is said to be regular if gl.dim $R < \infty$ or equivalently, if K.dim $R = \dim_{R/\Omega}(\Omega/\Omega^2)$. If R is not a local ring, then R is termed to be a regular ring if for every maximal ideal Ω, $Q_{R-\Omega}(R)$ is regular.

2.3. Theorem. Let R be a graded commutative Noetherian ring, then the following statements are equivalent :

1. R is a regular ring.
2. $Q_{R-P}(R)$ is regular for every graded prime ideal P of R.

Proof : The implication 1. ⇒ 2. is obvious.

2 ⇒ 1. If P' is an arbitrary prime ideal of R then the ring $Q_{R-(P')_g}(R)$ is regular i.e. gl.dim $Q_{R-(P')_g}(R) < \infty$. Since $Q_{R-P'}(R)$ is the localization of $S = Q_{R-(P')_g}(R)$ at the prime ideal generated by P' in it, it follows that we may assume that we have chosen R to be gr-local with gr.maximal ideal $(P')_g$. By hypothesis gl.dim $S = n < \infty$. Consider a finitely generated R-module M. By Proposition 1.2. we have that $\text{inj.dim}_S Q_{R-(P')_g}(M) \leq n$ and hence $\text{gr.inj dim}_R M \leq n$. Therefore gr.gl.dim $R < \infty$ and Corollary 1.7. yields gl.dim $R < \infty$, hence gl.dim $Q_{R-P'}(R) < \infty$.

Our following result shows that regular and positively graded rings are close to being polynomial rings over regular rings.

2.4. Proposition. Let R be a positively graded regular ring with a unique graded maximal ideal Ω, then R_0 is a regular local ring and $R \cong R_0[X_1,\dots,X_k]$, the X_i, $i=1\dots k$, being indeterminates over R_0, which are taken to be homogeneous elements of positive degree.

Proof : We proceed by induction on $n = K.\dim R$. If $n=o$ then $R=R_0$ is a field. If $n>o$, choose a nonzero homogeneous element x of positive degree in $\Omega-\Omega^2$ (if there is no such x then $R=R_0$ follows). Put $R_1 = R/(x)$, $\Omega_1 = \Omega/(x)$. Now R_1 is a graded regular ring and $K.\dim R_1 = n-1$.

The induction hypothesis yields : $(R_1)_o$ is a regular local ring and $R_1 \cong (R_1)_o[Y_1,\dots,Y_k]$ with $\deg Y_i > o$, $i=1\dots k$. Since $(x)\cap R_o = o$ it follows that $(R_1)_o \cong R_0$. So, if $K.\dim R_o = c$ then $d+k = n-1$. Choose representatives $U_1,\dots,U_k$ in R for the images of $Y_1,\dots,Y_k$ in R_1, and put $U_{k+1} = x$. One easily checks that $R = R_0[U_1,\dots,U_{k+1}]$. Therefore we have an epimorphic graded ring homomorphism : $\psi : R_0[X_1,\dots,X_{k+1}] \to R_0[U_1,\dots,U_{k+1}] = R$. However, the fact that $K.\dim R = k+d+1 = K.\dim R_0[X_1,\dots,X_{k+1}]$ yields that ψ is an isomorphism. □

2.5. Remarks. 1. If R is a positively graded regular ring then R_0 is regular. Indeed, if ω is a maximal ideal in R_0 then $\Omega = \omega \oplus R_+$ is a maximal graded ideal. Since the localization $Q_{R-\Omega}(R)$ is regular and also positively graded (note that it coincides with the graded ring of quotients $Q^g_{R-\Omega}(R)$), it follows from the foregoing that $(Q_{R-\Omega}(R))_0$ is regular. However $(Q_{R-\Omega}(R))_0 = Q_{R_0-\omega}(R_0)$. It follows that R_0 is regular too.

2. See also J. Matijevic [24]. If $R = \bigoplus_{i\geq o} R_i$ is Gorenstein, hence Cohen-Macaulay, then it is not necessarily true that R_0 is Gorenstein or C.M.. Indeed, put $T = k[X,Y]/(X^2,XY)$, where R is any field. Put $Q_1 = (Y)$, $Q_2 = (X)$, then the image of $Q_1 \cap Q_2$ in T is zero. Let W,V be indeterminates of degree 1 and put $S = T[W]$, $Q_1^e = Q_1S$, $Q_2^e = Q_2S + WS$ and $I = Q_1^e \cap Q_2^e$. Put $U = S/I$, $R = U[V]/(xV+yV,V^2)$ where x,y denote the images of X,Y resp. This construction yields a graded ring R with $K.\dim R = 1$ and $R_0 = T$. However T is not a C.M.-ring.

If Ω denotes the unique maximal graded ideal of R then $Q_{R-\Omega}(R)$ is Gorenstein.

III.3. Graded Rings and M-sequences

R is a commutative graded ring throughout this section. Such a ring R is said to be completely projective or is said to have property c.p. if for any graded ideal I and for any finite set of graded prime ideals $p_1,\dots,p_n$ with $h(I)\subset p_1\cup\dots\cup p_n$ with $h(I)\subset p_1\cup\dots\cup p_n$ it follows that I is contained in at least one p_i. It is not hard to verify that, if R is completely projective then so is $S^{-1}R$ for every multiplicatively closed subset S of h(R). It is also straightforward to check that epimorphic images of a completely projective ring have property c.p. too.

3.1. Lemma. Let R be a commutative graded ring such that all graded prime ideals of R are in Proj R, then R is completely projective.

Proof : Assume $I\not\subset p_i$, $1\leqslant i\leqslant n$ and assume that the p_i are not comparable one to another. If $n=1$ then $h(I)\not\subset p_1$ follows. We proceed by induction on n. The induction hypothesis yields $h(I\cap P_i)\not\subset \bigcup_{j\neq i} P_j$, $1\leqslant i\leqslant n$. Hence there is an $a_i\in h(I\cap P_i)$ such that $a_i\notin p_j$ for each $j\neq i$. Let $x_i=\prod_{j\neq i} a_j$. Clearly $x_i\in I$ and $x_i\in p_j$ for $j\neq i$ but $x_i\notin p_i$. Since no graded prime ideal of R contains R_+ is easy to see that we may choose $\deg x_i>o$ for each i, $1\leqslant i\leqslant n$. Put $d_i=\deg x_i$ and $d=d_1\dots d_n$, $y_i=x_i^{d/d_i}$. Then $\deg y_i=d$ and it is obvious that $y=y_1+\dots+y_n$ is a homogeneous element such that $y\in I$ and $y\notin P_1\cup\dots\cup P_n$. □

3.2. Examples. 1. If R is a positively graded ring such that R_o is a field then R is completely projective.

2. Let R be a gr.local ring with gr.maximal ideal Ω such that R/Ω is a graded division ring with non-trivial grading. Then R has property c.p. Indeed, if P is a graded prime ideal then $P\subset\Omega$. Since R/Ω is non-trivially graded, there is a homogeneous t of positive degree, $t\notin\Omega$, hence $t\notin P$.

3. Let $\mathcal{O}$ be a discrete valuation ring with maximal ideal ω generated by t say. Put $R=\mathcal{O}[x]\cong\mathcal{O}[X]/(tX)$ i.e. in R we have $tx=o$. Then (t,x) is a graded ideal and every

homogeneous element of (t,x) is a zero-divisor; however $t+x$ is not a zero-divisor. If $P_1,\dots,P_n$ are the prime ideals associated to R, then these are graded and clearly $h(I) \subset P_1 \cup \dots \cup P_n$ but $I \not\subset P_i$ for each i, $1 \leq i \leq n$. So $\mathcal{O}[x]$ does not have property c.p.

Let R be a commutative Noetherian ring and M a finitely generated R-module. A sequence $a_1,\dots,a_n$ in R is said to be an <u>M-sequence</u> if $(a_1,\dots,a_n)M \neq M$ and for each $i=1,\dots,n$, a_i is not an annihilator in the module $M/(a_1,\dots,a_{i-1})M$. If I is an ideal of R such that $IM \neq M$, then we shall denote by grade(I,M) the common length of all maximal M-sequences contained in I. As it was already mentioned in Section III.2, it is well known that grade(I,M) is the smallest number r such that $\mathrm{Ext}^r_R(R/I,M) \neq 0$.

<u>3.3. Corollary</u>. Let R be a commutative Noetherian ring which is graded and completely projective. Let $M \in R$-gr be a finitely generated object and I a graded ideal of R such that $IM \neq M$. Putting grade(I,M) $= m$, then there exists in I an M-sequence of length m, consisting of homogeneous elements. Moreover any M-sequence formed by homogeneous elements of I has the same length i.e. grade(I,M).

<u>Proof</u> : For $m = 0$ the assertion is clear. Let $m \geq 1$ and let $P_1,\dots,P_s$ be graded prime ideals associated to M. Since grade(I,M) > 0, $I \not\subset P_1 \cup \dots \cup P_s$ and thus because of Lemma 3. $h(I) \not\subset P_1 \cup \dots \cup P_s$. Hence there exists $f_1 \in h(I)$ such that $f_1 \not\in P_1 \cup \dots \cup P_s$, therefore f_1 is a non-annihilator of M. Let M_1 be the graded R-module M/f_1M, then grade(I,M) $= m-1$ and we apply the induction hypothesis. The second statement may also be proved using a similar induction argument. □

<u>3.4. Corollary</u>. Let R be a commutative Noetherian graded ring having property c.p. Let P be a prime ideal of R having height n. Then there exist homogeneous elements $a_1,\dots,a_n$ in R such that P is minimal over $Ra_1 + \dots + Ra_n$.

<u>Proof</u> : For $n = 0$ there is nothing left to prove, so we may assume $n > 0$ and proceed by induction on n. Let $\{Q_1,\dots,Q_k\}$ be the set of minimal prime ideals of R and as R is graded, each one of the Q_i, $i=1,\dots,k$, is graded. Since $\mathrm{ht}(P) \geq 1$, P is not contained in any Q_i and therefore, Lemma 3.1. entails $h(P) \not\subset Q_1 \cup \dots \cup Q_k$. Pick $a_1 \in h(P) - \bigcup_{i=1}^{k} Q_i$.

In R/Ra_1, P/Ra_1 is a graded prime ideal with $ht(P/Ra_1) \leq n-1$. The induction hypothesis yields that P/Ra_1 is minimal over $(\bar{a}_2,\dots,\bar{a}_n)$, where the $\bar{a}_i$, $i=2,\dots,n$, are homogeneous elements of R/Ra_1. Choosing representatives a_i for $\bar{a}_i$ in R it is obvious that P is minimal over $(a_1,\dots,a_n)$. □

Consider a gr.local ring R with maximal graded ideal Ω and suppose that R is completely projective. Let $M \in R$-gr have finite type. The ring $\bar{R} = R/\mathrm{Ann}_R M$ is a gr.-local ring with maximal graded ideal $\bar{\Omega} = \Omega/\mathrm{Ann}_R M$ and $\bar{R}$ has property c.p. Write n for the Krull dimension of M, then $n = \mathrm{K.dim}\ \bar{R} = ht(\bar{\Omega})$. By Corollary 3.4., there exist homogeneous elements $\bar{a}_1,\dots,\bar{a}_n \in \bar{\Omega}$ such that $\bar{\Omega}$ is minimal over $\bar{R}\bar{a}_1 + \dots + \bar{R}\bar{a}_n$ in $\bar{R}$. Our hypothesis yield that $M/a_1M + \dots + a_nM = M/\bar{a}_1M + \dots + \bar{a}_nM$. Furthermore $\bar{R}/\bar{R}\bar{a}_1 + \dots + \bar{R}\bar{a}_n$ is a gr.Artinian ring, hence $M/\bar{a}_1M + \dots + \bar{a}_nM$ is a gr. Artinian $(\bar{R}/\bar{R}\bar{a}_1 + \dots + \bar{R}\bar{a}_n)$-module. On the other hand we have that $\bar{R}/\bar{R}\bar{a}_1 + \dots + \bar{R}\bar{a}_n = R/\mathrm{Ann}_R M + Ra_1 + \dots + Ra_n$, thus it follows that $M/a_1M + \dots + a_nM$ is a gr.Artinian $(R/\mathrm{Ann}_R M + Ra_1 + \dots + Ra_n)$-module; consequently, the latter module actually is gr. Artinian in R-gr.

A system of homogeneous parameters for the module M with $\mathrm{K.dim}\ M = n$ is a set of elements $a_1,\dots,a_n \in h(\Omega)$ such that $M/a_1M + \dots + a_nM$ is Artinian in R-gr (hence of finite length in R-gr!).

3.5. Remarks. 1. As in the ungraded case one may establish that every M-system may be included in a system of homogeneous parameters for M.

2. If R is positively graded and such that R_0 is a field then there exists a system of homogeneous parameters for R_+. However if R_0 is not a field but a local ring with maximal ideal ω then the graded ideal $\Omega = \omega + R_+$ may not have a system of homogeneous parameters, as the following example shows.

3.6. Example. Let A be a discrete valuation ring with maximal ideal ω and uniform parameter t. Let $R = A[X]/(tX)$; then R is Noetherian and positively graded. Moreover R is a gr. local ring with Krull dimension 1. One easily verifies that there cannot exist a homogeneous $a_1 \in (\omega + R_+)/(tX)$ such that $\{a_1\}$ is a system of homogeneous parameters.

In the light of the foregoing observations we now suppose that the graded Noetherian ring R has the form $R_0 \oplus R_1 \oplus \dots$, where R_0 is a field. Then R is gr.local with gr.maximal ideal R_+.

<u>3.7. Proposition</u>. Let $M \in R$-gr have finite type and let $\{a_1,\dots,a_n\} \in h(R_+)$. A necessary and sufficient condition for $a_1,\dots,a_n$ to be a system of homogeneous parameters for M is that M has finite type as an object of $R_0[a_1,\dots,a_n]$-gr , while n is minimal as such.

<u>Proof</u> : If $a_1,\dots,a_n$ is a system of parameters for M then $\overline{M} = M/a_1M + \dots + a_nM$ is Artinian in R-gr. Since $\overline{M}$ has finite type its grading is left limited. On the other hand, since $\overline{M}$ is Artinian in R-gr and since R is positively graded, the grading of $\overline{M}$ is also right limited. Consequently $\overline{M}$ has to be a finite dimensional R_0-vector space. Nakayama's lemma then yields that M is finitely generated as a $R_0[a_1,\dots,a_n]$-module. Conversely, if we start from the assumption that M is a finitely generated $R_0[a_1,\dots,a_n]$-module, then $M/a_1M + \dots + a_nM$ is a finite dimensional R_0-vector space, thus Artinian in R-gr. Next let us check minimality of n as such. Indeed, if $\{b_1,\dots,b_m\}$ is a set of homogeneous elements such that M is a finitely generated $R_0[b_1,\dots,b_m]$-module then $M/b_1M + \dots$ $\dots + b_mM$ is Artinian in R-gr, i.e. of finite length too. But then $K.\dim M \leq m$ excludes the possibility $m < n$. □

<u>3.8. Proposition</u>. Let $M \in R$-gr have finite type. If $a_1,\dots,a_n$ is a system of homogeneous parameters for M then the elements $a_1,\dots,a_n$ are algebraically independent over R_0.

<u>Proof</u> : Letting $a_1,\dots,a_n$ be such a system of parameters, consider the ring $S = R_0[X_1,\dots,X_n]$ with gradation defined by putting $\deg X_i = \deg a_i$, $i = 1,\dots,n$. Specializing $X_i \to a_i$ defines a surjective ring homomorphism of degree o, $\varphi : R_0[X_1,\dots,X_n] \to R_0[a_1,\dots,a_n]$. We have that $K.\dim R_0[a_1,\dots,a_n] = K.\dim M = n$. However if $\operatorname{Ker} \varphi \neq o$, then the Krull dimension should have dropped, so φ is monic and an isomorphism. □

<u>3.9. Proposition</u>. Let $M \in R$-gr be finitely generated. The following statements are equivalent :

1. M is a Cohen-Macaulay module.

2. If $a_1,\dots,a_n$ is a system of homogeneous parameters for M then M is a free (graded free) $R_o[a_1,\dots,a_n]$-module.
3. Every system of homogeneous parameters for M is an M-sequence.

Proof : 1 ⇒ 2. Put $S = R_o[a_1,\dots,a_n]$. By Proposition 3.8., S is a regular ring with K.dim S = n. Since M is a Cohen-Macaulay module, grade(R_+,M) = K.dim M = n, and from Corollary 3.3. there exists an M-sequence $b_1,b_2,\dots,b_n$ with $b \in h(R)$ $i = 1,\dots,n$. For $i = 1,\dots,n$, put $M_{(i)} = M/b_1M + \dots + b_iM$ and put $M_{(o)} = M$. Considering the fact that S is a regular ring, we may derive from the exactness of $o \to M_{(i)} \xrightarrow{m_{i+1}} M_{(i)} \to M_{(i+1)} \to o$, (where m_{i+1} denotes multiplication by b_{i+1}) that : $(\star)$ $p.\dim_S M_{(i+1)} = 1 + p.\dim_S M_{(i)}$.
On the other hand, $M_{(n)} = M/b_1M + \dots + b_nM$ is an S-module of finite length. Therefore $M_{(n)}$ is a finite dimensional R_o-vectorspace. Because S is gr.local we have that $p.\dim_S M_{(n)} = p.\dim_S R_o = gl.\dim S = n$. However from $(\star)$ it follows that $p.\dim_S M_n = n + p.\dim_S M_{(o)}$. Now $p.\dim_S M_{(n)} = n$ yields $p.\dim_S M_{(o)} = p.\dim_S M = o$, consequently M is a free S-module.
2 ⇒ 3. As well as 3 ⇒ 1 are rather easy. □

In the sequel we aim to give some applications of the theory just expounded to the study of filtered rings and modules.

3.10. Proposition. Let R be a filtered ring with exhaustive filtration FR and let $M \in R$-filt have the exhaustive filtration FM. Suppose that all submodules of M are closed in the topology defined by the given filtration. Let $r_1,\dots,r_n$ be elements of R and denote by a_i, $i = 1,\dots,n$, the image of r_i in the graded ring G(R). If the homogeneous elements $a_1,\dots,a_n$ form a regular G(M)-sequence then $r_1,\dots,r_n$ form a regular M-sequence. Moreover for each k, $1 \leq k \leq n$ we have :

$$G(M/r_1M + \dots + r_k M) \cong G(M)/a_1G(M) + \dots + a_k G(M).$$

Proof : We proceed by induction on n.
If n = 1; let $x \in M$, $x \neq o$. For some $p \in \mathbb{Z}$, $x \in F_pM$ and $x \notin F_{p-1}M$. Since $x_p \neq o$ (x_p is the image of x in G(M), for notation etc., see Section I.4), we have that :

$o \neq a_1 x_p = (r_1 x)_{p+p(1)}$, where $p(1)$ is defined as follows, $r_1 \in F_{p(1)}R$ and $r_1 \notin F_{p(1)-1} R$. Therefore $r_1 x \neq o$ and r_1 is M-regular. On the other hand we have : $G(M/r_1M) \cong G(M)/G(r_1M)$ and one sees that $G(r_1M) = a_1G(M)$. Assume now that our assertion is valid for $n = k$. Since $r_1M + \ldots + r_kM$ is closed it follows that $M/r_1M + \ldots + r_kM$ is separated in the quotient filtration. The fact that a_{k+1} is not an annihilator of $G(M)/a_1G(M) + \ldots + a_kG(M)$, together with Proposition I.4.1. yields that r_{k+1} is not an annihilator of $M/r_1M + \ldots$ $\ldots + r_kM$. Using the induction hypothesis in the case $n = k$ and $n = 1$ we obtain :

$$G(M)/a_1G(M) + \ldots + a_{k+1}\, G(M) = (G(M)/a_1G(M) + \ldots + a_kG(M))/(a_{k+1}G(M)/a_1G(M) + \ldots + a_kG(M))$$

$$= G(M/r_1M + \ldots + r_kM)/a_{k+1}G(M/r_1M + \ldots + r_kM)$$

$$= G(M/r_1M + \ldots + r_kM)/G(r_1M + \ldots + r_{k+1}\, M/r_1M + \ldots + r_kM) = G(M/r_1M + \ldots + r_{k+1}M). \quad \square$$

<u>3.7. Theorem</u>. Let A be a Noetherian local ring with maximal ideal ω and suppose that FA is a filtration satisfying $F_iA = o$, $i > o$, $F_oA = A$, $F_{-1}A = \omega$, and F_n, $n < o$ defines the ω-adic topology on A. Let $R = G(A) = \bigoplus_{i \leq o} F_iA/F_{i-1}A$, $\Omega = \bigoplus_{i \leq -1} F_iA/F_{i-1}A$ the canonical maximal graded ideal of R. Then :

1. If $Q_{R-\Omega}(R)$ is a Cohen-Macaulay ring then A is C.M.
2. If $Q_{R-\Omega}(R)$ is a Gorenstein ring then A is Gorenstein.

<u>Proof</u> : 1. If $Q_{R-\Omega}(R)$ is C.M. then by Theorem 2.1. it results that R is C.M. Now R is completely projective, hence we may select homogeneous elements $a_1, \ldots, a_n \in R$ which form an R-sequence with $n = ht(\Omega) = K.\dim R$. Proposition 3.6. entails that $\text{grade}(\omega, A) = K.\dim A$, i.e. A is a Cohen-Macaulay ring.

2. Since $Q_{R-\Omega}(R)$ is a Gorenstein ring then R is a Gorenstein ring. The argumentation of the proof in 1. allows to reduce the proof of 2. to the case where A is an Artinian local ring. We need :

<u>Sublemma</u>. If A is an Artinian local ring then equivalently :

1. A is a Gorenstein ring.
2. $(o : \omega) = \{a \in A,\ a\omega = o\}$ is a minimal ideal.
3. There exists $z \in A$, $z \neq o$ such that for every $x \neq o$ in A, there exists an element $y \in A$ such that $xy = z$.

Proof of the sublemma : $1 \Leftrightarrow 2$, cf. [1]. $3 \Rightarrow 2$ Obvious. $2 \Rightarrow 3$ Take $z \in (o:\omega)$, $z \neq o$ and take $o \neq x \in A$. If $x\omega = o$ then $x \in (o:\omega)$ and by 2. there is an $y \in A$ such that $xy = z$. If $x\omega \neq o$ then pick $x_1 \in \omega$ such that $xx_1 \neq o$. Consider the ideal $xx_1\omega$, then obviously $x\omega \supset xx_1\omega$ since equality would entail $\omega = x_1\omega$, a contradiction if $\omega \neq o$. Repeating this construction and using the descending chain condition we find that there exist elements $x_1, \dots, x_k$ such that $xx_1x_2 \dots x_k \neq o$ and $xx_1x_2 \dots x_k\omega = o$. Thus $xx_1 \dots x_k \in (o:\omega)$ and minimality of $(o:\omega)$ yields that there is a $y \in A$ such that $xx_1x_2 \dots x_ky = z$.

Returning to the proof of the theorem, we have already established that $R = G(A)$ is a Gorenstein ring which is a local ring with maximal ideal ω. So we may apply the sublemma, i.e. suppose that a is a homogeneous element of R satisfying the condition 3 of the sublemma. (Note that the fact that Ω is a graded maximal ideal of R plus the fact that $(o:\omega)$ is also graded allows indeed to find homogeneous elements satisfying 3.) Pick $\alpha \in A$, say $\alpha \in F_pA$, such that $\alpha_p = a$. Then for every $o \neq \gamma \in A$, say $\gamma \in F_qA$, there exists a homogeneous $b \in R$ such that $\gamma_q b = a$. Let $\beta \in A$ be such that β maps to b in $G(A)$. Then $\gamma_q b = a$ yields that $\gamma\beta - \alpha \in F_{p-1}A$. Taking $\gamma \in F_{p-1}A$ we find that b has positive degree i.e. $b = o$ but this is impossible, hence $F_{p-1}A = o$. Consequently for arbitrary $\gamma \neq o \in A$ and α and β as above, it follows that $\gamma\beta - \alpha = o$. This states exactly that α satisfies condition 3 of the sublemma and therefore A is a Gorenstein ring. □

REFERENCES

1. H. Bass, On the ubiquity of Gorenstein rings, Math. Z. (1963), 8-28.
2. N. Bourbaki, Algèbre, chap. 2-8, Hermann, Paris (1962).
3. N. Bourbaki, Algèbre commutative, chap. 1-7, Hermann, Paris (1965).
4. K.S. Brown, E. Dror, The Artin-Rees property and homology, Israel J.Math. vol.22. (1975), 93-109.
5. R. Barattero, Alcune caratterizzazioni degli anelli H. Macaulay ed. H. Gorenstein, Le Matematiche vol XXIX (1974), 304-319.
6. H. Cartan, S. Eilenberg, Homological Algebra, Princeton University Press (1956).
7. W.L. Chow, On unimixedness theorems, Amer. J. Math. 86, (1964), 779-822.
8. R. Fossum, H.B. Foxby, The Category of Graded Modules, Math. Scandinavica (1974), vol 35 no.2.
9. P. Gabriel, Des Catègories Abéliennes, Bull. Soc. Math. France, (1963), 90, 324-448.
10. J. Golan, Localization of Noncommutative Rings, Marcel Dekker (1975), New York.
11. A.W. Goldie, Semi-prime rings with maximum conditions, Proc. London Math. Soc. 10 (1960), 201-220.
12. R. Gordon, J.C. Robson, Krull Dimension, Amer. Math. Soc. Memoirs (1973), vol. 133.
13. A. Grothendieck, Sur quelques points d'algèbre homologique, Tohoku Math.J. 2 (1957) no.9, 119-183).
14. M. Hochster, L.J. Ratcliff, Five theorems on Macaulay rings, Pacific. J. of Math. 44 (1973), 147-172.
15. B. Iversen, Noetherian Graded Modules I., Aarhus Universitet, preprint No.29, 1972.
16. G. Krause, On the Krull-dimension of left Noetherian left Matlis rings, Math. Z. 118 (1970), 207-214.
17. I. Kaplansky, Commutative Rings, Allyn and Bacon, (1970), Boston Mass.
18. I.D. Ion, C. Năstăsescu, Anneaux gradués sémi-simples, Rev. Roum. Math. 4 (1978) 57
19. A.V. Jategoankar, Left Principal Ideal Rings, Lect. Notes in Math. vol. 123, Springer Verlag 1970, Berlin.
20. A.V. Jategoankar, Jacobson's Conjecture and Modules over Fully Bounded Noetherian Rings, J. of Algebra, vol. 30, 1-3 (1974), 103-121.
21. D. Lazard, Autour de la platitude, Bull. Soc. Math. France, 97 (1969), 81-128.
22. E. Matlis, Injective Modules over Noetherian Rings, Pacific J. Math. vol. 8 (1958), 511-528.
23. J. Matijevic, P. Roberts, A conjecture of Nagata on graded Cohen-Macaulay rings, J. Math. Kyoto Univ. vol. 14 (No. 1 (1974).

24. J. Matijevic, Three local conditions on a graded ring, Trans. Am. Math. Soc., vol. 205 (1975), 275-284.

25. M. Nagata, Local Rings, Interscience (1962), New York.

26. M. Nagata, Some questions on Cohen-Macaulay rings, J. Math. Kyoto Univ. 13 (1973), 123-128.

27. Y. Nouazé, P. Gabriel, Idéaux premiers de l'algèbre enveloppante d'une algèbre de Lie nilpotente, J. of Algebre, vol. 6 (1967), 77-99.

28. C. Năstăsescu, Anneaux et Modules gradués, Rev. Roum. Math. no. 7 (1967), 911-931.

29. C. Năstăsescu, Décompositions primaires dans les anneaux noethériens à gauche, vol. XXXIII-13, Symposia Matematica.

30. F. Van Oystaeyen, On Graded Rings and Modules of Quotients, Comm. in Algebra. 6(18), 1923-1959 (1978).

31. F. Van Oystaeyen, Graded and Non-Graded Birational Algebras, Ring Theory, Proceedings of the 1977 Antwerp Conference, Lect. Notes in Math., vol. 40, Marcel Dekker, 1978, New York.

32. F. Van Oystaeyen, Prime Spectra in Non-commutative Algebra, Lect. Notes in Math. vol. 444 (1975) Springer Verlag, Berlin.

33. F. Van Oystaeyen, A. Verschoren, Reflectors and Localization. Application to Sheaf Theory., Lect Notes in Math., vol. 41 Marcel Dekker Inc. 1979 , New York.

34. Rentschler, Gabriel,
C.R. Acad. Sci. Paris 265 (1967), 712-715.

35. G. Sjödin, On filtered modules and their associated graded modules, Math. Scand. 33 (1973), 229-249.

36. B. Stenström, Rings and Modules of Quotients, Lect. Notes in Math. vol. 237, Springer Verlag (1971), Berlin.

37. R. Sharp, Gorenstein Modules, Math. Z. 115, (1970), 117-139.

38. O. Zariski, P. Samuel, Commutative Algebra vol. X, Van Nostrand, (1960), New York.

Subject Index

Vol. 580: C. Castaing and M. Valadier, Convex Analysis and Measurable Multifunctions. VIII, 278 pages. 1977.

Vol. 581: Séminaire de Probabilités XI, Université de Strasbourg. Proceedings 1975/1976. Edité par C. Dellacherie, P. A. Meyer et M. Weil. VI, 574 pages. 1977.

Vol. 582: J. M. G. Fell, Induced Representations and Banach *-Algebraic Bundles. IV, 349 pages. 1977.

Vol. 583: W. Hirsch, C. C. Pugh and M. Shub, Invariant Manifolds. IV, 149 pages. 1977.

Vol. 584: C. Brezinski, Accélération de la Convergence en Analyse Numérique. IV, 313 pages. 1977.

Vol. 585: T. A. Springer, Invariant Theory. VI, 112 pages. 1977.

Vol. 586: Séminaire d'Algèbre Paul Dubreil, Paris 1975–1976 (29ème Année). Edited by M. P. Malliavin. VI, 188 pages. 1977.

Vol. 587: Non-Commutative Harmonic Analysis. Proceedings 1976. Edited by J. Carmona and M. Vergne. IV, 240 pages. 1977.

Vol. 588: P. Molino, Théorie des G-Structures: Le Problème d'Equivalence. VI, 163 pages. 1977.

Vol. 589: Cohomologie l-adique et Fonctions L. Séminaire de Géométrie Algébrique du Bois-Marie 1965–66, SGA 5. Edité par L. Illusie. XII, 484 pages. 1977.

Vol. 590: H. Matsumoto, Analyse Harmonique dans les Systèmes de Tits Bornologiques de Type Affine. IV, 219 pages. 1977.

Vol. 591: G. A. Anderson, Surgery with Coefficients. VIII, 157 pages. 1977.

Vol. 592: D. Voigt, Induzierte Darstellungen in der Theorie der endlichen, algebraischen Gruppen. V, 413 Seiten. 1977.

Vol. 593: K. Barbey and H. König, Abstract Analytic Function Theory and Hardy Algebras. VIII, 260 pages. 1977.

Vol. 594: Singular Perturbations and Boundary Layer Theory, Lyon 1976. Edited by C. M. Brauner, B. Gay, and J. Mathieu. VIII, 539 pages. 1977.

Vol. 595: W. Hazod, Stetige Faltungshalbgruppen von Wahrscheinlichkeitsmaßen und erzeugende Distributionen. XIII, 157 Seiten. 1977.

Vol. 596: K. Deimling, Ordinary Differential Equations in Banach Spaces. VI, 137 pages. 1977.

Vol. 597: Geometry and Topology, Rio de Janeiro, July 1976. Proceedings. Edited by J. Palis and M. do Carmo. VI, 866 pages. 1977.

Vol. 598: J. Hoffmann-Jørgensen, T. M. Liggett et J. Neveu, Ecole d'Eté de Probabilités de Saint-Flour VI – 1976. Edité par P.-L. Hennequin. XII, 447 pages. 1977.

Vol. 599: Complex Analysis, Kentucky 1976. Proceedings. Edited by J. D. Buckholtz and T. J. Suffridge. X, 159 pages. 1977.

Vol. 600: W. Stoll, Value Distribution on Parabolic Spaces. VIII, 216 pages. 1977.

Vol. 601: Modular Functions of one Variable V, Bonn 1976. Proceedings. Edited by J.-P. Serre and D. B. Zagier. VI, 294 pages. 1977.

Vol. 602: J. P. Brezin, Harmonic Analysis on Compact Solvmanifolds. VIII, 179 pages. 1977.

Vol. 603: B. Moishezon, Complex Surfaces and Connected Sums of Complex Projective Planes. IV, 234 pages. 1977.

Vol. 604: Banach Spaces of Analytic Functions, Kent, Ohio 1976. Proceedings. Edited by J. Baker, C. Cleaver and Joseph Diestel. VI, 141 pages. 1977.

Vol. 605: Sario et al., Classification Theory of Riemannian Manifolds. XX, 498 pages. 1977.

Vol. 606: Mathematical Aspects of Finite Element Methods. Proceedings 1975. Edited by I. Galligani and E. Magenes. VI, 362 pages. 1977.

Vol. 607: M. Métivier, Reelle und Vektorwertige Quasimartingale und die Theorie der Stochastischen Integration. X, 310 Seiten. 1977.

Vol. 608: Bigard et al., Groupes et Anneaux Réticulés. XIV, 334 pages. 1977.

Vol. 609: General Topology and Its Relations to Modern Analysis and Algebra IV. Proceedings 1976. Edited by J. Novák. XVIII, 225 pages. 1977.

Vol. 610: G. Jensen, Higher Order Contact of Submanifolds of Homogeneous Spaces. XII, 154 pages. 1977.

Vol. 611: M. Makkai and G. E. Reyes, First Order Categorical Logic. VIII, 301 pages. 1977.

Vol. 612: E. M. Kleinberg, Infinitary Combinatorics and the Axiom of Determinateness. VIII, 150 pages. 1977.

Vol. 613: E. Behrends et al., L^p-Structure in Real Banach Spaces. X, 108 pages. 1977.

Vol. 614: H. Yanagihara, Theory of Hopf Algebras Attached to Group Schemes. VIII, 308 pages. 1977.

Vol. 615: Turbulence Seminar, Proceedings 1976/77. Edited by P. Bernard and T. Ratiu. VI, 155 pages. 1977.

Vol. 616: Abelian Group Theory, 2nd New Mexico State University Conference, 1976. Proceedings. Edited by D. Arnold, R. Hunter and E. Walker. X, 423 pages. 1977.

Vol. 617: K. J. Devlin, The Axiom of Constructibility: A Guide for the Mathematician. VIII, 96 pages. 1977.

Vol. 618: I. I. Hirschman, Jr. and D. E. Hughes, Extreme Eigen Values of Toeplitz Operators. VI, 145 pages. 1977.

Vol. 619: Set Theory and Hierarchy Theory V, Bierutowice 1976. Edited by A. Lachlan, M. Srebrny, and A. Zarach. VIII, 358 pages. 1977.

Vol. 620: H. Popp, Moduli Theory and Classification Theory of Algebraic Varieties. VIII, 189 pages. 1977.

Vol. 621: Kauffman et al., The Deficiency Index Problem. VI, 112 pages. 1977.

Vol. 622: Combinatorial Mathematics V, Melbourne 1976. Proceedings. Edited by C. Little. VIII, 213 pages. 1977.

Vol. 623: I. Erdelyi and R. Lange, Spectral Decompositions on Banach Spaces. VIII, 122 pages. 1977.

Vol. 624: Y. Guivarc'h et al., Marches Aléatoires sur les Groupes de Lie. VIII, 292 pages. 1977.

Vol. 625: J. P. Alexander et al., Odd Order Group Actions and Witt Classification of Innerproducts. IV, 202 pages. 1977.

Vol. 626: Number Theory Day, New York 1976. Proceedings. Edited by M. B. Nathanson. VI, 241 pages. 1977.

Vol. 627: Modular Functions of One Variable VI, Bonn 1976. Proceedings. Edited by J.-P. Serre and D. B. Zagier. VI, 339 pages. 1977.

Vol. 628: H. J. Baues, Obstruction Theory on the Homotopy Classification of Maps. XII, 387 pages. 1977.

Vol. 629: W. A. Coppel, Dichotomies in Stability Theory. VI, 98 pages. 1978.

Vol. 630: Numerical Analysis, Proceedings, Biennial Conference, Dundee 1977. Edited by G. A. Watson. XII, 199 pages. 1978.

Vol. 631: Numerical Treatment of Differential Equations. Proceedings 1976. Edited by R. Bulirsch, R. D. Grigorieff, and J. Schröder. X, 219 pages. 1978.

Vol. 632: J.-F. Boutot, Schéma de Picard Local. X, 165 pages. 1978.

Vol. 633: N. R. Coleff and M. E. Herrera, Les Courants Résiduels Associés à une Forme Méromorphe. X, 211 pages. 1978.

Vol. 634: H. Kurke et al., Die Approximationseigenschaft lokaler Ringe. IV, 204 Seiten. 1978.

Vol. 635: T. Y. Lam, Serre's Conjecture. XVI, 227 pages. 1978.

Vol. 636: Journées de Statistique des Processus Stochastiques, Grenoble 1977, Proceedings. Edité par Didier Dacunha-Castelle et Bernard Van Cutsem. VII, 202 pages. 1978.

Vol. 637: W. B. Jurkat, Meromorphe Differentialgleichungen. VII, 194 Seiten. 1978.

Vol. 638: P. Shanahan, The Atiyah-Singer Index Theorem, An Introduction. V, 224 pages. 1978.

Vol. 639: N. Adasch et al., Topological Vector Spaces. V, 125 pages. 1978.